Personal Currency:
The Future of Money

Personal Currency: The Future of Money

By Roger Colburn & Alex Riegelmann

Published by Lionhead Press

Lionhead Press
Colorado Springs, Colorado
www.lionheadpress.com

Although the authors and publisher have made every effort to ensure the accuracy and completeness of information contained in this work, we assume no responsibility for errors, inaccuracies, omissions, or any inconsistency herein. Any slights of people, places, or organizations is unintentional.

Printed in the United States of America

ISBN-13: 978-0-9892285-2-7

First Edition

Ordering Information:
Quantity sales. Special discounts are available on quantity purchases by corporations, associations, and others. For details, contact the publisher at the address above.
Orders by U.S. trade bookstores and wholesalers. Please contact Lighting Source: Tel: 1-800-509-4156; or visit www.lightningsource. com.

Chapters

Introduction

The human world, as we know it, is made up of systems. We have a system for everything; from fishing and farming to construction or programming.

We create laws, which comprise the "justice" system. A network of vehicles and people comprising "transportation" systems. Doctors, nurses, hospitals, researchers, chemists, and manufacturing plants that make up "medical" systems. Systems give us the order and security we crave as humans.

Systems, as we know them today, are man-made. We design them ourselves and they tend to continuously evolve over time. Sometimes our systems go on as usual for decades, without fundamental shifts; other times, they experience major overhauls or breakthroughs that are nearly always controversial. With the rise of incredible new technologies in recent history, these changes are more rapidly happening to our institutions and our way of life. In our remarkably connected world, new breakthroughs travel so fast that they often cannot be stopped or controlled; much like how Uber ran away with the long-established taxi industry, or how cell phones almost completely supplanted corded phones.

This is what happens when innovations are allowed to permeate and spread through our worldwide culture naturally, without interference. There are many institutions, however, that are not allowed to evolve and change as new ideas appear.

The single largest institution in the world, in terms of power, is the financial system. It is one which everyone uses, and also has an effect on every other human system (whether directly or indirectly). But whereas once money was allowed to change and take different forms depending on the needs of people using it, now it is tightly controlled, and has remained more or less unchanged since 1971.

Money is our tool for exchange, and for many of us, it is a force to live by. Money has the power to give life and growth to entire nations of people, and it has the power to kill with its lack. The best thing about money is that it allows us to trade openly and flexibly, and bridges gaps between otherwise dissimilar businesses.

Yet money is one of the least understood things that we use every day. Ask most people about how banking works, and they won't be able to tell you anything beyond the basics.

"You put your money in the bank. Alternatively, you can borrow money from them, and pay interest on it in addition to the sum you borrowed." Sounds simple enough. Until we go a little deeper.

Don't forget that the bank pays depositors interest for keeping their money there. Why? Because they lend that money out to others.

Wait a minute—what if I come back to get all my money? Will it all be gone?

No, it won't. In fact, your money wasn't even used—it was copied.

Again, why? Well, the answer to this—and many other questions about money—are not common knowledge. But they are essential to understanding how money will change in the future.

Because our economies change, money must also change. We must accept that with new, unprecedented technologies, we will also experience a shift in thinking about the things we use, and the people we interact with, every day. Money is just a tool we use for transacting, which means that the transactions are what matter. Transactions are, after all, just elements of relationships. Good transactions are good for relationships; they conjure connections and feelings of mutual faith between people. Bad transactions, which carry connotations of betrayal and derogatory terms like "ripoff," are bad for relationships, breeding mistrust and hostility.

That's why money is so important. Think of all the transactions you make every day. Whether or not money is involved, they are important to you, and others. They frame your relationships and your family. You make dinner, your wife vacuums. Kate does the current payroll, and you cover the expense report. Tony welds the parts on while you mix the paint. A gallon of milk and small talk at the supermarket.

A new TV at the electronics store. Coffee in the morning. If everyone feels it's fair, all smile at the end, satisfied. Ultimately this satisfaction stems from consideration—some kind of currency, money or not, was provided by everyone, and it felt equal and honest.

Yet when the technical subject of "economy" comes up, we usually gloss over an essential issue: what is currency, and how does it work? Who decides how it works, and why? Are we doing it right, or could money be better?

We're going to take an honest look at money, from a basic view all the way up to the way that modern banking works, explained in the clearest terms possible. Then we'll talk about what it means, why it will change, and how.

...Because it will. In fact, money is long overdue for a change. Beyond new "economic models," "banking laws," and the usual conspiracy theories, there are new and exciting ideas for the future of money that go beyond the usual "isms" (like Capitalism, Marxism, Socialism, Anarchism, etc.) and actually harness technologies and conventions that have existed for years.

Rather than profit, future money will foster a focus on people, and their business—what they are "busy" doing.

Life is business, and business is life. And soon we'll change the way we do it.

A digital revolution is coming to money.

Part I: How Currency Works

"Neither a borrower nor a lender be."

William Shakespeare

CHAPTER ONE

MONEY

What is money? What makes it a medium of
exchange? Money has changed profoundly over its
history, corresponding to how we think about it.

A PRE-MONEY WORLD

We tend not to think of a time before money. Currency
seems timeless and ubiquitous. Something so fundamental to
human life that we must have always used it... but we didn't. In
fact, money is a relatively new invention, albeit an important one.

Before money, life was simpler, but without it, civilization
(as we know it) was not possible. There were fewer people in
the world, and they lived in small communities and associations
called "tribes." As far as we know, the idea of "capitalism" didn't
exist until money did (hence the use of "capital," meaning
"money"). These earlier humans practiced something called a Gift
Economy.

Gift economy was notably observed in native North
American societies. Visiting Europeans found their generosity
and apparent lack of ownership concepts very strange. In the
Iroquois and Huron tribes of New England, common culture
revolved around relationships, rather than materials. Social
capital (social wealth) was considered the most important and
valuable thing a person could have. This was smart—in a hunter-
gatherer society, or one with primitive agriculture and almost no
government, people had to form close communities in order to
survive. It was safer and ensured that work could be completed
quickly.

Because the Iroquois and Huron tribes practiced
agriculture, they actually built and maintained small towns, and
in order to do so, they needed workers to farm the land and build

structures. Yet instead of paying people for this labor, or having a system of currency in which people could receive money in exchange for their services or goods, it was simply established that everyone could take whatever they needed, when they needed it, as long as they performed some kind of service to the rest of the community. In other words, if you work, you eat.

Interestingly, in this society, because there was no concept of individual ownership, there was no concept of theft (how can you steal what's yours?). Additionally, economic competition was non-existent.

The degree to which the community was wealthy was the degree to which you were wealthy. This apparently motivated people to work unlike anything else; in fact, among the Iroquois, work ethic and virtue were synonymous. Interdependence, rather than independence, was what everyone strove for, both communally and individually; since quality of life improved with cooperation.

According to archaeological evidence, it appears that this was the first type of economy; from when humans first left their original homelands (in Africa, as contemporary archaeology suggests) until they invented money, around 3000 BCE; roughly five thousand years before the 21st century.

Among societies practicing gift economics, there appears to be a strong sense of community and cooperation, as well as a general sense of belonging and individual motivation to make positive contributions. One area of Western society that we still see gift economics every day is within the traditional family. In families, there is typically a strong drive (especially by the parents) to provide for others within the group, with no expectation of reciprocation. This is especially evident in family-based businesses like farms, where all members are expected to contribute in some way, the consequences of failing to do so being social rather than economic (i.e. If you don't do your chores, you might be shunned, or Dad will be angry. But ideally, no one will force you to starve to death).

Many principles of gift economics seem socially superior to modern capitalism. Yet there were good reasons for the invention of money; of capital, without which, as mentioned, civilization (as we know it) could not have arrived.

FROM COINS TO CREDIT

Wealth. Capital. Credit. Currency. It has many names, and to some it means little, and to some, it is everything. It is a way to store value that seems to become more and more complicated as time goes on—from its humble beginnings as simple, crude chunks of metal with symbols in them to the electronic, changeable, and volatile virtual currencies of today. Despite its crucial role in human life, and our constant exposure thereto, few people actually understand what money is. What does it do? What makes it valuable? Why do nations rise and fall because of it? Unless you know what money is, you cannot understand the incredible transformation that money is about to undergo in our society.

Money is the tool which allows our society to move. It is a lubricant which allows the various pieces and functions to operate. The cogs of the magnificent machine which makes up human society are comprised of businesses. Businesses run everything from the donut shop down the street to the mail in your mailbox—and yes, even those things which do not seem like businesses, like the police force, the organizations that maintain the park or the national forest, the military, or the local library, are all engaged in everything that's required to make them businesses—they require money to operate, they employ individuals to do so, and they provide a product or service.

Value is the totally intangible idea that something is useful, and therefore worthy of human effort to acquire. Value is what money is made of—in our minds. Physically, money today is made of things like paper, or cotton, or various metals. Various metals, which, even in the past when rarer metals made up money, tended to be largely useless—otherwise they wouldn't have been wasted on tokens only representing an idea.

The entire idea behind "money" as we know it today is as a storage of value. It is precisely because it takes a long time and a lot of effort to move cows, sheep, wood or wheat, that we invented money. It represents the inherent value of other things; things that actually serve a purpose, so that we wouldn't always have to trade directly in order to get something we need.

With money, products could travel further, businesses could produce more, and people more easily found food, shelter, and temperature security. Money also enabled something else, something that transformed human society forever—the commercialization of art.

In the ancient world, you would be hard-pressed to find anyone willing to trade some bread for a beautiful realist painting of a sunset. That's because the painting itself had no use—it was only nice to look at, and "nice to look at" doesn't plow fields or feed cattle. This is the reason that some of the earliest commercial art (art you pay for) was on pottery, shields, or weapons. As an example, a pot is used to store liquid, grain, and other things—very useful. Why get one pot over another? Perhaps one has pretty things on it, and another doesn't. Hence the decoration. Why decorate your armor, shields, and weapons? Identification on the battlefield. Very useful... life-and-death useful.

But when money entered the fold, and value became detached from goods directly, everything changed. People experienced more excess and more freedom—money had flexibility, in that after it was used to purchase whatever a person/family needed, the leftover value could be freely used for, well, anything. This allowed people the luxuries of things like art without having to be a king or warlord.

It also allowed more commercial momentum. With that excess capital, you don't have to buy art. You can invest in something that might potentially make you more money. Are you a dairy farmer? Sell enough milk to buy a bull, so you can breed more cattle. In the pre-money world, the farmer across the way wouldn't trade with you if he didn't need what you have. If he was also a dairy farmer, then there are not very many ways for you to business with him. But with money, you don't need to have something he needs. Because you share a mutually, widely recognized form of currency, you can trade just fine. You give him some of your cash, he gives you a bull, and in a few years, you have way more cows, and therefore, a far greater output of value (which is good for the rest of your community, as well as you). You not only gain money, you gain security.

Lose a few cows to disease? Because of the commercial momentum that money gave you, you're probably going to be fine. Money was a fine invention.

The first monies were transitory and changeable. People initially used very common commodities like barley as money, which is why the term "shekel" came to be used for money in the Middle East—"shekel" was a unit of weight, used for barley.

More abstract things like seashells began to arise as money sometime later. Eventually, around 600 BC, coins made of electrum (gold & silver alloy) were used in Anatolia (modern-day Turkey). This practice spread, since the production of coins could be centrally controlled by a king, government, or a business, directly. It allowed rulers to control money, and therefore, economic power, in their respective societies.

With the rise of huge empires like Rome, money became complicated. Because it was either based in a real commodity or in some kind of precious metal, transporting wealth took a large amount of effort and a degree of danger. Caravans carrying wealth had to be protected, sometimes heavily, and not even that always worked. Along with the difficulty of transport, every king, and therefore every kingdom, had a different currency, and the perceived values of these currencies changed like the wind.

This is what gave rise to banks. A bank allows you to place your wealth somewhere that it will be protected, and you can come get it whenever you need it, rather than having to store it yourself (and assume the risk of doing so). Beginning in the Renaissance era in Europe, banks would take in your gold or silver and then give you a receipt, which was a token that allowed you to redeem the gold/silver at any time. These tokens, which represented the real value of what you put into the bank, came to be their own form of exchange; people began trading these directly, rather than making the extra trip to get their precious gold/silver out of the bank first. As long as you had the bank note, you could retrieve it yourself, so it simplified transactions and made them a little bit safer. This is what gave rise to paper money.

When governments caught on to this banking practice, they formed central banks. A central bank is exactly like a private bank—except it is controlled directly by a government and its representatives/rulers. In many ways, by the time the Renaissance came around, central banks had already existed for a long time.

Kingdoms had long been producing their own coins and hoarding gold, silver, and whatever else was valuable. They needed these things to wage war, which was usually their most important activity. Without the ability to wage war, a government is essentially useless, since it is unable to prevent its people from being subjugated by... someone else.

Central banks, however, were what gave rise to the first centralized paper money. This helped people trade with each other within nations—it reduced the risk of using bank notes from different banks. What if someone paid you with a bank note for thirty pounds of silver, and when you went to the bank to redeem it, found that the bank was robbed? Not only are you not getting any silver, but the note you have is useless. It's not good anywhere else. Yet with a central bank, your deposits would be (in theory) protected by the military of an entire nation. Your bank's security would be a regular army. This is not only good for your deposits, it is also good for the people you might be trading with—they are going to feel much better about accepting your paper money (bank notes) than if they were from some smaller local bank.

The usage of representative money became more and more simple as time went on. With the rise of centralized paper money, the smaller banks adapted by accepting deposits of the bank notes themselves. It was less risky than having the actual gold or silver stored somewhere. As time went on and these practices became widespread, the gold and silver stored up in the banks became less and less important. Bankers realized that people only needed to believe that the precious metal was there—which led to the next big stage for money: fractional reserve banking.

SILVER, THEN GOLD, THEN... NOTHING

To reprise, reserve banking means that if you have a note promising a certain weight/amount of something and you bring it to the bank, you get that thing. So, the bank should always have 100% of that thing that people have deposited. But what if people never come and get their stuff? Their gold, their silver? Or only a small percentage do, in the banker's experience.

In some cases, only about 10% of people would come back to the banks to get their deposit at any given time. So the banks, in order to make greater profits from interest payments, realized that they could lend out the money that wasn't theirs. As long as they kept a certain percentage, or a reserve of the gold, silver, or paper money on hand, it was almost guaranteed that they could lend out the rest, and have it paid back with interest. Hence the term "fractional reserve," which describes the modern form of banking. This practice began during the Renaissance era and has continued ever since, becoming more and more complicated through the years as the valuation of money moved away from gold and silver.

In 1971, United States President Richard Nixon finally ordered that the U.S. dollar would no longer be redeemable for gold. This changed very little for most people, since dollars (federal reserve notes) were accepted across the world at face value. And yet, in principle, this changed the US dollar's value substantially, and had far-reaching consequences. Because without a reserve requirement based on something real, like gold, the potential *amount* of U.S. currency became limitless. It was no longer based on anything—which meant it was "good" for purchases regardless of the amount of gold, silver, rubies, tortillas, or anything else stored by the central bank. Thus the system of money we have today was born: in which the value of money is based on "fiat," or a governmental order that it is worth something, and nothing else.

Yet you wouldn't know it, would you? The Bank of England still prints "I promise to pay the bearer on demand the sum of ... pounds" on its currency notes. Wouldn't logic seem to say that if you have that currency note, you have that currency? Why does the note itself say that it is just a promise? This is the confusing reality of modern currency. It claims to have intrinsic value, or to represent the value of something else, but in reality, does no such thing. Statements like "I promise to pay the bearer" or "federal reserve note" are empty and meaningless. What would the bearer be paid? A federal reserve of what? More "notes?"

Modern money isn't really... anything. Yet, the world uses it as if it actually does store value, because the fact that we agree it does is *all that matters*. And that is where much of the genius, as well as the problems with money, have come from.

CHAPTER TWO

VALUE

Value is an entirely human concept that we often think of as objective—but it's far from it.

NOT A LONG TIME AGO...

It was early on January 27th, 2014, when CONCORD, the enforcers of interplanetary government, suddenly appeared in system B-R5RB, and demanded that the bill for sovereignty of the system be paid by HAVOC, the corporation which laid claim to this particular region of space. Unable to do so, or perhaps due to some clerical error, the "rent" on the system proceeded unpaid, which prompted CONCORD to revoke HAVOC's sovereignty over the system.

HAVOC's failure to pay the sovereignty tax would not have been such an issue (it could have been easily renewed), if this particular system were not an extremely important strategic foothold in the ongoing war between the Pandemic Legion (the alliance which owned the HAVOC corporation) and the CFC Alliance and their allies; most notably, Russia.

Because, you see, B-R5RB was the staging point for the entire Legion battle fleet. Taking the system would mean a decisive strategic victory for the CFC Alliance and Russian command. So, not wanting to miss the opportunity, the coalition mobilized every super-capital and capital-class ship they could, and rushed through the warpgates armed to the teeth.

They began deploying their best ships; the enormous, feared, and coveted Titans, in force. Dozens of these enormous, kilometers-long doomsday ships began to arrive, and quickly assembled into battle groups to attack the floating space station, which, if successful, would give them the ability to control the traffic, resources, and weapon platforms.

If they captured this control station, the Legion's fleets would be trapped in B-R5RB, rendering them sitting ducks for the coalition to destroy.

Bluish shields and glistening metal twinkled with starlight as the impossibly large fleet moved toward their objective with impunity.

PL was aware of the error, however. A similar mobilization immediately took place on their side; since this system was so important, they would not risk losing it, no matter the cost. Many of their most powerful ships were already present, although notably, their numbers were weaker than the gigantic attack force now appearing in their sector. CFC and Russian commands quickly moved to attack the smaller force—but this day, a quick victory was not to be had.

The PL force, with little hope of defeating their adversary, fearlessly baited and delayed the enemy fleet long enough for the rest of their allied super-capital ships to arrive—which ensured that the largest battle ever seen in the galaxy could ensue. Over 6,000 pilots participated, which lasted a whole 12 hours without stopping. 55 alliances were mobilized, and 717 corporations participated.

A stalemate dragged on, during which, neither side was able to gain a foothold. Awe-inspiring blasts from planet-cracking superweapons crisscrossed through space like white-hot comets, leaving trails of sparkling molten metal where smaller ships were vaporized on contact.

Finally, the tables turned when the CEO of the Northern Coalition, Vince Draken, took command of the entire Legion and allied forces. He ordered that all primary fire be focused on Sort Dragon, the titan from which the Russian fleet admiral commanded, in an attempt to decapitate his enemy's force.

Despite the combined power of the PL's impetuous navy, the Russian fleet and their allies smartly concentrated their combat repair and shield support systems enough to prevent Sort Dragon from being destroyed. By the time the titan had taken any significant damage, the Legion had lost 5 titans of their own. Over the next hours, the Legion was slowly pushed back, and the combined Russian and CFC Alliance navies continued to pour more and more of their forces, all that they had, into the conflict; until, all at once, after tens of thousands of ships and their pilots were scattered into ionized dust, all the beams, guns, and missiles fell silent. PL had withdrawn, and the exhausted pilots of the CFC Alliance were victorious. In B-R5RB, broken hulls the size of small countries formed slow-orbiting clouds of man-made debris that still linger today.

The cost? 11 trillion Inter-Stellar-Kredits, or ISK. The equivalent of 300-500 thousand US dollars.

What was just described isn't from some science fiction fantasy book or movie. It was a totally unscripted event that occurred in a video game called EVE Online, a space-based online jaunt through a fictional universe where faster-than-light travel and interstellar warfare are everyday realities. The reason it's significant is the price tag: 11 trillion ISK, which is really worth $300 thousand. Fake money with real value. Which means it is not really fake—as long as people *believe* it has value, it does.

THE IDEA OF VALUE

"Value" is the idea that something gives worth, or contains worth, for human life. We assign "value" to many things. We think of tools as valuable. Food is valuable—it keeps us alive by feeding our bodies. But we have other needs: mental, emotional, spiritual needs. We assign value to those things as well. For example, films and TV shows serve a need for something intangible: for entertainment.

"Man does not live by bread alone," a famous man once said, and this is certainly true—we require more than just calories, water, and air to function. According to Abraham Maslow, American psychologist and author of the popular "Hierarchy of Needs," food and basic survival are the least of the essential human concerns—security, belonging, esteem, and accomplishment are all more important, and fulfilling, than these. We therefore assign great value to the things that give us fulfillment in the higher areas, such as entertainment.

Maslow said the "physiological needs" were the first and most basic need of human beings, without which, any desire for the others hugely diminishes (if not disappears). When these are met, we have the need for safety—to be secure in our physical selves, to have a place to sleep away from nocturnal predators, abuse, natural disaster, disease, or whatever else can cause bodily harm. Next comes love and belonging, which includes both sexual and non-sexual intimacy. Next is esteem, which is the feeling of having respect from others. This spurs confidence in people, and leads to a desire for the final need: self-actualization, or the realization of one's mission in life, or one's full potential, put simply. As we achieve the higher levels of need, we naturally place less value on things that meet the lower levels. We pay the highest prices, in currency, time, effort, value for the things at the "top" of the hierarchy, and the lowest values for things at the bottom.

Creatures which are simpler than human beings have more basic needs, which are therefore simpler to meet. In fact, human needs are, arguably, the most complex of any creature on the planet. We have complex and intricate emotional needs, especially, which is part of the reason we have so many different vocations, professions, types of art, and products. Whenever a need arises in the human consciousness, someone rises up to offer a solution for that need, by creating a product that's tangible or intangible, depending on the need.

For example, according to popular legend, the Africans who discovered coffee did not perceive it as having value until they noticed their goats and other animals becoming energetic after eating the coffee cherries. They tried the cherries themselves, and quickly found that their need and desire for energy was satisfied. Coffee plants and their seeds (or "beans") quickly spread through the Arab world, then to Europe. Without the discovery that coffee meets a certain human need, it never would have been considered valuable.

Yet the process we use to determine value is slightly more complicated than that. The amount of value placed on something tends to be directly correlated to that thing's perceived utility. Hence our preoccupation with price, which is designed to be a direct quantity of value. The basic implication of price is that the higher the price for something, the greater the sacrifice that must be made to acquire it, and therefore the more valuable it must be.

DIAMONDS OVER WATER

Except it's not so simple. Price is a paradox, because value itself is subjective. Adam Smith, classic economist, described this with a famous illustration: that despite diamonds' total uselessness for human survival and water's total necessity for survival, diamonds are priced much, much higher than water. According to this situation, it would seem that price and therefore value have nothing to do with utility. Apparently, price, and therefore value, correlates more with effort. If water were difficult to obtain, like diamonds, there would be less of it available, and we would consider it more valuable.

Therefore, money, which is what we use to store value, is used as a measure of usefulness, which operates as a hierarchy. The more completely that something meets a need, the more valuable it is—multiplied by the level of the need (physiological needs being the lowest multiplier; existential or self-actualizing needs being the highest). This is important for understanding what makes money itself valuable—or really, what determines its psychological value.

There's even more to it. Supply is another factor in value. When supplies of something are low, even something typically easy to obtain, the price (psychological value) of it goes way up. Price is how we give measurable value to things, and yet, price is extremely subjective. This is part of the reason that the modern monetary systems are so complicated. The price of houses, for example, fluctuates based on factors that have little to do with the cost of building materials. It is the market, or an amalgamation of perceived value, which controls this. Pretty mind-warping.

The complexity of housing prices aside, it is clear that value is a totally subjective psychological construction (as is its measure, money). Therefore, our best bet for understanding money and value is understanding people, or at least what motivates them.

In the very real present, for some people, not-starving is their biggest day-to-day concern. At this level of poverty, several of the aforementioned needs of the hierarchy are not met. Security needs are likely not met, as impoverished people don't have enough money to meet expenses, let alone save, and physiological needs are likely also suffering. Since the meeting of these needs is dependent on money, the perceived value of money by people in this situation is proportionally much higher than it is to people with an excess of money. But because money has become centralized, its value is wholly dependent on the whims of the government that creates it, in an attempt to "objectivize" something naturally subjective. Thus, the money we use today; the money that we use to transact with each other, always involves the forced psychological construction of a "third party"—the government that created it.

This was originally a way to keep people safe, and to help the economy. It was for a common good. Yet as time goes on, we are finding ourselves beset by more monetary problems and far fewer monetary conveniences, as the world gets larger—and the economy, like an enormous growing vine, gets more complex.

CHAPTER THREE

MODERN MONEY CHALLENGES

**Our monetary system is in a unique "deadlock"
that prevents it from evolving according to the
needs of society, as it did in the past.**

CONVENIENCE WITH CAVEATS

Money solved a big problem for human society—before it, wealth could not travel very far or very fast. Money brought flexibility to our lives by allowing transactions between many different types of people in many different places. It simplifies our economy at the individual level.

Money, indeed, solves many of the problems inherent barter—but it also introduces its own problems. For one, part of what makes money valuable in our minds is the fact that there isn't enough of it. It is permanently "scarce." This means that someone is always left without enough.

In addition, the way money is created is a perpetual cycle of debt which can never be completed. Because most money has moved into digital realms today, only a small amount of the actual money supply is represented in cash. This has accelerated money's movement, to be sure, but it has also accelerated money's creation. US dollars are created when a bank approves a loan. Conventional logic would say that a bank would make loans from the money they already had stored up—but this is actually not the case. Banks create the money (on "paper," or in the modern world, inside a computer system) then lend it to the borrower. When the loan is repaid to the central bank, the money is essentially destroyed.

However, the money created is due back to the institution which created it. But if all money is created this way, then where is the money required to pay off all the debt?

Well, It doesn't exist.

In addition to the money borrowed, interest must also be paid on said funds. Yet if all money was originally borrowed, and new money must be created by borrowing it, then where is the money to pay the interest?

That also doesn't exist. Not even on paper, or in a computer system. It's just the way the system is set up; there will never be enough to pay off the debt. Therefore all funds are actually part of a chain of debt that ends with someone's default on said debt. Which is a little scary, isn't it? Default is built-in to our monetary system.

Not that it was designed expressly to cause trouble. Defaults are kept to a minimum (though never really eliminated) as long as the amount of available goods and services (the economy, essentially) continue to increase as time goes on. Perpetual growth is necessary for the monetary system to work, and this is why the US economy is considered to be suffering when growth drops below a certain rate, rather than stopping altogether.

The major problem (which the overseers of our monetary system are certainly aware of) is that growth cannot go on forever. We live on a planet of finite resources, and grade-school physics say that such a system will exhaust various natural resources, if not only the available open space to live on. Sustainability is the mark of efficiency, and our monetary system is actually just the opposite. It is wholly unsustainable.

This is, of course, the big-picture problem with money. There are many more problems with money, particularly at the individual level.

POVERTY IN DOLLARS

Conventional money is designed expressly to return to the place it was created. It is, after all, a loan. Thus, modern money isn't really meant to be a storage of value. In fact, people saving too much was one of the reasons for ending the gold standard; financial moguls noticed that spending was what spurred an economy to grow. In fact, consumer spending makes up 70% of the US economy today—meaning that saving is actually bad for the economy as a whole.

"Economy" comes from a Koine Greek word, *oikonomia*, which means "household management." When we talk about "the economy" we are really speaking of the management of a country's resources. A successful management of resources (or a successful management of anything) would be one which is efficient and beneficial to those who use those resources. An economy which does not benefit the people using it is therefore ineffective. And yet, we find that the unit of value that controls our resources (money) is mostly accumulated by one very small section of the population, and the rest are dependent on that population for the resources they need.

And indeed, money is a poor store of real wealth. Stored dollars continually become less valuable as they inflate, or devalue, because more dollars are being created relative to the goods and services being produced in the US. This is called "inflation," and it has been perpetual since 1955. The reason for inflation? Debt.

Since the 1950s, and especially the '60s, the United States has been accumulating more debt each year. Armed conflicts like the Vietnam and Korean Wars were especially costly, which was one of the reasons foreign investors were losing confidence in the dollar. During the Vietnam conflict, many foreign powers began to suspect that the United States was lending more money than they possessed in gold reserves, and they were right. West Germany was the first to leave the international gold standard entirely, and were followed by Switzerland shortly thereafter. Others began to demand huge sums of gold, such as France, who withdrew $191 million.

This led Richard Nixon, president, to end the US dollar's redeemability in gold and effectively keep US gold reserves from being drained by other nations. Ever since, the dollar has not been tied to any commodity, which makes it easy to print. The Federal Reserve has been doing so, in order to pay fixed-amount debts and to meet the demand for dollars overseas and domestically, in the US itself.

Complex systems of international debt and repayment aside, the US dollar is not an instrument of wealth (nor is any other currency); it is an instrument and of debt, and of managing debt. It does not allow people to create wealth; only acquire it from somebody else, and even that must return to its originator some day. This fundamental problem is preventing business—the basic engine of society—to function at its maximum potential.

The answer, through, is not to return to a commodity-based system, such as the gold standard. This system was proven to be ineffective for widespread use, since gold is not producible. It, or any other single commodity, will always end up in the control of some kind of institution, as the gold of the United States is.

It's not so much where the value originates from, but how it can be used, that makes a currency an effective representation of value. Since the US dollar is a poor storage of wealth, and is designed to travel, like a boomerang, back to its source, and is subject to an extremely complex system of control, it is an effective representation of value for only a few people, not for the whole group that is forced to use it. There is wealth, or value, in many places, and especially in people, which is not represented by dollars. For example, in one week, a carpenter may produce twenty wooden chairs; yet until someone buys those chairs from him, which in practice, are inherently valuable (there's nothing wrong with them, they took work to produce, they serve a purpose, etc.), he has no way to represent the value of those chairs. Because he is not a bank, he cannot create a redeemable currency and spend it, using the chairs as collateral.

But what if he could?

WEALTH IN SELF

Our cultural concept of wealth is so tied up in currency, we think of them as one and the same—yet value isn't currency, or vice-versa. In truth, it is only represented by currency. Value actually exists, intrinsically (by itself) in many things: useful tools, sentimental objects, copyright/virtual property (such as an online photo album), and, perhaps most obviously, relationships. Of course, what makes things like these things valuable is a matter of philosophy, but suffice to say for our reasons in this book, they are valuable because people agree they are (much like currency itself). Yet we can't use relationships as currency, as an asset to trade with. Nor can we use our unsold writings, or paintings, or film work, or even the music that we are not making, not because we don't have the talent, but because we are too busy working a job in order to procure dollars.

Wait... what is preventing us from using these things as currency? Is the carpenter, who possesses 20 solid, well-made chairs (and the potential to produce 20 more next week!) going to starve despite his incredible talent and ability for making furniture, if nobody buys them? The answer is yes, but only because he, and you, and everyone, is forced to use dollars.

The answer to this flaw in centralized, "fiat" currency, is not a return to an older way of thinking, and doing. What has been tried before was dropped when something new and more appropriate was found. Similarly, we are in need of a new and more appropriate system of money, and mindset thereof, that takes the lessons of the past and uses them along with present and future innovations. The world has changed, and is changing, at an incredible rate, and it is time to embrace, in all areas of life, new ways of thinking and believing.

There is an incredible, a staggering amount of real wealth in the goods, services, and most importantly, in the time and energy of the people living in the United States, and yet, our monetary system, the thing we have all chosen to store this wealth with, is ill-proportioned, inefficient, and causes huge misappropriations of energy and people.

How do we release this wealth? How can we re-appropriate or augment our monetary system to allow more wealth, and thus, more prosperity; a better way to manage our resources? There must be a better way to do money, and this is what spurred the genesis of alternate currencies.

CHAPTER FOUR

MONEY ALTERNATIVES

Alternative currencies are as old as currency itself. In fact, there are many already circulating around the world, serving many colorful uses and adaptations to unique economic conditions.

ONE MAN'S TRASH

Before the American Civil War, around 1,600 corporations and private banks created their own paper currencies, which were payable on demand in gold and silver coin. The US government only created coins at this time, which accounted for only 4% of the money supply. It was during the Civil War that the first standardized US dollars appeared in the form of "federal reserve notes," which were nicknamed greenbacks. They were designed to pay the debts and expenses of the war, which were substantial.

The US Constitution stated that only gold and silver, in the form of coins, could be considered official US currency, and this was because the paper money created by Congress during the Revolutionary War, called "continentals," became worthless by the end of the war, and produced economic problems (for one, the British were able to counterfeit them on a huge scale). This soured the US government to paper money. But the concept of fiat money was too good to stay away (since it was so convenient for paying and keeping track of debts). It was re-instated during America's next great war, and was there to stay.

Before huge-scale wars, commerce and exchange in general were far more flexible and dependent on the agreements made by the parties involved.

The larger that regular armies became, the more the governments supporting them needed money, not only to pay soldiers, but for equipment. War is a black hole for money—it causes a huge outflow of funds without producing anything... except death and destruction. In other words, it is the worst application of resources, the use of economy, that we know of. The rationale for these expenses is that they are for defense; whatever the reason for guns and bombs, they must be paid for somehow, and debt is a great way to do this with essentially no limits. After all, every nation claims an absolute justifiable right to win the war. Cost should not matter.

As Roman politician Marcus Tullius Cicero said, "The sinews of war are infinite money." Crippling debt and a devaluing, defunct currency were contributing reasons for the fall of Rome. Roman currency of the time (the denarius) lacked substitutes, as it was considered the "imperial standard," and was even used in non-Roman territories as money (much like the US dollar today). This centralization and total reliance on the denarius was at first beneficial to Rome, as it encouraged wealth to stay within the empire, but later it became a huge detriment as the money supply, relative to Rome's need to fight wars, dwindled, and the new money they minted contained less and less actual value.

After the Civil War, the United States made the "greenback" the official and primary currency for the entire country. This was, in part, to provide the financial resources needed to rebuild after the devastation of war, and also looking to the future, as the nation expanded west. While the greenback was intended at first to be a temporary measure, there were so many in circulation that by the Civil War's end, it was easier to simply solidify the currency than it was to deprecate it. So it was renamed "United States Dollar" and was thenceforth the country's primary currency.

This, however, like many state-sponsored economic interventions, had some unintended consequences that reached far into the future.

FORCED DEPENDENCE

Without alternatives to the dollar, Americans became economically dependent on it. It became so essential to business that if it were to become defunct, there would be nothing to replace it. In today's world, bartering might make a partial comeback, but as Chapter 1 illustrated, it wouldn't be flexible enough to serve the economic needs of the modern world. Civilization is built on money. Complex supply chains would quickly break down in an all-barter system. No, with the absence of the dollar, the US government would have to introduce a new currency very quickly, or find a way to revalue the dollar. Even then, the damage would have been done. Since currency is about belief, a loss of faith in the issuer of the currency is worse than a loss of faith in a currency itself—it makes it very difficult for that issuer to regain the trust needed to distribute currency again.

This is why alternative forms of exchange were a successful economic buffer in the past. Anything can be an alternate currency, including the official currency of another nation. When the United States was a new country, the "Spanish Dollar" (8 Spanish Reales) was acceptable for the payment of debts, as well as the British Pound Sterling. The individual colonies also printed their own currencies, which were used alongside each other in commerce.

This, whether intentionally or not, was a self-regulating system of currency. Currency values and efficacy were determined by popular opinion rather than government decree, and so some currencies would rise or fall in the marketplace purely due to their ability to serve a useful purpose, or more accurately, whether people believed they did.

The government of the United States did produce official coins that were used alongside everything else. Therefore, these varieties of currency weren't really alternate, but complementary—they were used along with the currencies already in circulation. For individuals, this meant a more secure storage of wealth, believe it or not; if confidence was lost in a bank's (or a country's) notes, you were likely to have different "promises" (money) from other people (or organizations) that were still good.

At that time, most currencies, even paper notes printed by banks, were redeemable, or had direct value in, silver and gold. This limited currency creation to large institutions who were able to procure these metals.

There are actually numerous complementary currencies operating around the world today. None can truly replace the US dollar or other centralized currencies like Euros or Yen, but they do allow some economic freedom from these official currencies.

BerkShares - This is a local currency that is used throughout the Berkshires region of Massachusetts. This currency's value is pegged directly to the US dollar—1 BerkShare is worth 1 dollar. However, people purchase the currency at 95 cents each; this 5% discount is an immediate incentive to use it, as it is still used at full dollar value. This system is designed to keep value inside the local economy; the more dollars that are converted into BerkShares, the more money is essentially "trapped" in the Berkshires region. This is probably one of the simplest (and most effective) local complementary currencies used in the United States.

Ithaca HOUR - This is also a local currency, but uses a different method of valuation from most—it is denominated as one hour of "work" which is generally accepted to be worth about $10. This has been the longest currently used alternative currency in the United States, with several million dollars equivalent of goods and services having been traded using the "HOUR." It is used almost exclusively in Ithaca, New York.

LETS - Local Exchange Trading Systems. These are systems of local credit clearing which are non-profit, democratically organized, and designed to keep wealth within communities; specifically, between businesses. Inside a LETS network, people earn credit by doing various jobs or functions of their regular jobs (usually, again, within businesses) and use this credit within the network with other businesses or individuals. Transactions are publicly and centrally recorded to ensure transparency and continued confidence in the network. Some LETS use their own currencies; some do not, and prefer to keep credit exchanges moneyless.

Crypto-Currency - BitCoin, Litecoin, and others. These are a relatively new invention which are created and stored exclusively in computers. Crypto-currency is designed to be totally decentralized, in that no individual or group can influence or change its rules, use, or value. These currencies are also "capped" in value; only a certain amount of them can ever be created, which simulates the scarcity of precious metals (like in a commodity-based currency system). BitCoin was the first crypto-currency, and each new crypto-currency has since been based on it.

M-Pesa - This currency has taken on a remarkable following in southern and eastern Africa, particularly Kenya. It was originally conceived as a way to pay cell phone bills and micro-financed loans via a mobile phone, but it quickly grew to become much more than that. It allows users to deposit and withdraw money from retail outlets, transfer money between users, pay bills, purchase cell phone airtime, and pay for goods without any kind of cash required. In areas where using cash can be dangerous or unwieldy, this makes for safer commerce and generally more convenient purchasing of goods and services. M-Pesa continues to grow throughout the world, most recently to India and Egypt. M-Pesa is the most-used mobile money system in the world— around 25% of Kenya's gross national product goes through M-Pesa. Because it is so robust and widely used, M-Pesa has been under continuous attack by banking institutions since 2008.

The problem with all of these alternate currencies is that they are not very widely used; or at least, they are only used, and thus designed, for a small section of the population. There are also many online barter networks, but these also are utilized by a small number of people at any given time—and not as a reliable replacement for money.

The US dollar continues to be ubiquitously used throughout most of the world as a reserve currency (as of this writing, it makes up about 61% of all reserves by itself), meaning that although there may be different national currencies, most are reliant on the dollar for their own value and the backing of their "promise" as a form of legitimate money.

These currencies all have innovative ideas and address different problems, and typically work well for the people who initially designed and agreed to use them. Alternate currencies are a very innovative way to reshape life, and society (albeit on a small scale) without using violence—because, after all, businesses are the "cogs" and basic components of civilization. Change how businesses work, and society changes along with them (since businesses *are* society).

Why Have Alternate Currency?

If the dollar is so ubiquitous and convenient, why not continue to use it? Why bother creating and pushing a totally new currency? The local benefits of an alternate or complementary currency are obviously notable, yet they don't work in all situations. What about international trade, or the fact that most larger companies have supply chains which stretch from city to city, and state to state? A common medium of exchange was one of the main reasons for the invention and common adoption of money in the first place.

One of the biggest reasons that alternate currencies have emerged in recent years is the public distrust revolving around the dollar, and the fractional-reserve banking system which is used to create it. Many people, understandably so, feel uncomfortable using money that perpetuates a cycle of impossible debt and ambiguous valuation. This was much of the motivation behind the ill-fated "Liberty Dollar," a gold-backed alternate currency. The creators of these "dollars," an organization called "National Organization for the Repeal of the Federal Reserve and Internal Revenue Code" (or NORFED) were declared criminals and the money itself was declared counterfeit currency. Perhaps one of the major reasons for the government's legal aggression against the Liberty Dollar and NORFED was their blatant hostility toward the Federal Reserve System. NORFED also attracted radical government dissenters.

Regardless of NORFED's mistakes, their short-lived popularity showed that many people believe in commerce on terms that the parties involved agree upon; not on the terms that a government or other organization sets for them.

Having ultra-flexible currency which allows value to travel along agreed-upon and mutually beneficial paths is another reason people turn to alternate currencies. Currency is a contract—and being able to define that contract builds trust between the contract holders and their reputations in the "economy" in general; the reputation of a business can mean the difference between incredible success and total failure, after all. This, historically, gave depth to economies, which did not run the risk of having a total fallout if a national currency lost its efficacy.

But today? An international loss of confidence in the national currency would be absolutely, unequivocally, catastrophic.

A "Run" on the Dollar

If a bank's customers lose confidence in that bank around the same time, they go back to the bank and begin to demand their funds. This is what we call a "run" on the bank, and it is a very bad situation for the institution. Because our fractional reserve system only requires the bank to have 10% of the money it has lent out on hand at any given time, then if more than 10% of customers show up demanding their deposits, the bank is forced to default.

A default is a declaration that a debt or other obligation cannot be paid. Default typically results in a seizure of an individual's or organization's assets to pay the debt, but in the modern world, banks do not actually contain any assets.

Therefore, when you walk into a bank in the United States, there is a plaque somewhere in plain view that states: each depositor insured to at least $250,000, then the acronym "FDIC" in big letters. This federal institution, the Federal Deposit

Insurance Company, was created in 1933 to insure bank deposits, after the financial crisis of 1929, when there was a massive run on many banks throughout the country. The FDIC prevents personal and widespread economic disaster by insuring bank deposits in case of a run.

But there is no safeguard against a run on the dollar itself. This would be an international event on a global scale, and would have significant repercussions for nearly every community of humans on Earth.

China is the largest holder of "US treasury securities," (bonds) which are basically notes that guarantee a government payback. These bonds take a certain amount of time to "mature" (they cannot be redeemed until a certain time). These bonds are sold at a discount, so that a profit is made after maturity. Usually, treasury bonds have very long maturity periods, measured in decades, but this encourages holders to re-sell them, meaning they can circulate between countries.

Many of China's treasury bonds have matured, so they would be able to sell them if they wanted to. If everyone decided to sell their stocks of treasury bonds, or even just the largest three did (As of this writing, China, Japan, and Belgium, in that order), it would cause the value of the US dollar to plummet. This is because the market would suddenly flood with tons of bonds, which would make them less valuable. Just as if you found a massive stockpile of diamonds and sold them all, the price of diamonds would go down, since they would no longer be as hard to find as they used to be.

This would cause layoffs, as prices on everything increase, and employers have to cut costs. Everyone's money would become less and less valuable, and a default on US debt would mean a huge decrease in exports and imports, meaning that many commodities that the US relies on other countries for would become prohibitively expensive as well, exacerbating the problem. Untenable prices would mean civil unrest, and to prevent or counteract this, the United States government would be forced to either revalue the dollar or introduce a new currency altogether. In any case, serious economic damage would have occurred, with many people losing investments and many business deals going up in smoke.

Between the World Wars in Weimar Germany, such a scenario famously occurred. Because Germany was forced to pay huge reparations (debts) to the Allied powers, they began to simply print more money to buy other currencies (with which they paid the reparations). This inflated the German Mark, but since they faced strict deadlines on which to pay, they had no other choice. The price of a single loaf of bread topped 3 billion marks before inflation was halted by issuing a new currency.

Unfortunately, this happened a few times during the 20th century; notably, to Argentina. In 1989, a governmental election coincided with a World Bank recall of loans (which Argentina could not pay), and inflation reached a peak of 12,000% in 1989. When their currency became worthless, citizens simply raided grocery stores in force, stockpiling supplies, and rioters ruled the streets. While society did not break down completely before a new currency was issued, the Argentina inflation crisis was a good example of how important currency is to the operation of business, and therefore of everyday life for all people.

When people no longer felt that they had a way to legitimately exchange, they resorted to crime, out of perceived necessity—because they neither possessed nor knew of any other way to get what they needed, despite having jobs and contributing positions in society.

Most economists consider a worldwide hyperinflation of the US dollar to be unlikely (it is more likely that the dollar will simply be replaced, gradually, by the Euro or Chinese Yuan as the world's primary reserve currency). The United States is the best economic friend of China, Japan, and the other investors; it would hurt them just as much to "dump" the dollar. Yet this still illustrates the fundamental reality of the dollar: it is an instrument for commerce which the people doing the commerce have no control over.

CHAPTER FIVE

FREEDOM THROUGH BUSINESS

**Currency is an agreement; why not use it as such?
More flexible media of exchange means more
exchange can take place. More exchange makes
everyone involved better off.**

YOUR PERSONAL BUSINESS

Each of us, everyone, has a business. We all have something that we are busy doing. By contrast, businesses also take the form of amalgamations of individual human drives and gifts, and these are often called "corporations" or "companies." They are no different, just larger, more powerful, and capable of doing more. They are still comprised of people, without whose "business," they would not exist.

The idea for money first spawned, in part, from the need to store value and to exchange with others; other businesses. We come from a past in which it was far more common to trade directly, or to improvise a type of currency, than it was to use a national currency. Using systems like these, modern civilization as we know it grew up.

Businesses are vital, indispensable actually, to society. Businesses are what produce, and do, everything. There are certain businesses, though, that siphon value from others: governments. When we use money invented by a government, we are making exchanges based on terms introduced by that government. The Euro, for example, is created by the European Central Bank, which decides the interest rates at which it can be borrowed, under which circumstances it can be used, and who can use it.

Remember, governments are businesses too—they provide goods and services, and they employ people to do so. Yet their goal, their mission, is not to thrive by creating more wealth and expanding their operations. A government's mission is to provide national defense; to protect their people from both foreign enemies, and to enforce the morals which the people agree upon (through law). But should money be controlled by such an entity? Is it necessary for them to provide the currency that everyone *must* use?

Yes and no. It is legal to create and use alternate media for exchange. You could even look at something like the US Dollar as a form of "scrip"—or "company currency" issued by the business we call the United States Federal Reserve, and its primary shareholder (read: owner), the US Government. Any business could create a similar form of exchange and attempt to circulate it, as long as they were not attempting to counterfeit the dollar or any denomination thereof.

Imagine if a major search engine created a currency called "SearchCoin" or something to that effect. Using SearchCoins, they might pay their employees, or perhaps use the coins to pay bills to other businesses. These "Coins," therefore, might have some kind of redeemability for something the search giant provides, like server space. Let's say 1 SearchCoin is worth 1 megabyte of server storage; therefore, if you had 1,000 coins, you could redeem them—that is, turn them in to the search giant, removing them from circulation—and in return, you receive 1 gigabyte (1,000 megabytes) of storage on the search giant servers (or perhaps any of their affiliated/owned services). This is similar to commodity currency (like the dollar under the Gold Standard, under which you could bring dollars to any bank and redeem their worth in gold). These SearchCoins could circulate through society, being used for all kinds of purchases, until they reached a person who actually wants/needs the search giant server space.

The reason it could work as currency, for any purchase, is the general agreement among US citizens that the search giant server space has value, and therefore, the SearchCoin has value. The idea is relatively simple, actually, and is exactly how alternative currencies work (where they exist).

The US dollar, and other central bank currencies, are conventionally similar. Dollars are actually "redeemable" for Treasury Bonds (like the ones China buys, as explained in Chapter 4). Anyone can turn in their dollars to the US Central Bank in exchange for Treasury Bonds (which are promises to pay you more dollars at some future date). These bonds are essentially instruments of debt that allow the buyer to collect interest.

What if every company, every business, made their own money; and it was redeemable for their goods, services, or whatever they happened to provide for society (excluding debt, which is not actually a product)? There would be potentially thousands of currencies, millions even, circulating throughout society, each one readily redeemable for a good or service.

Such a system might naturally self-regulate. Only businesses that people trust would have valuable currencies, which would promote honesty and the importance of reputation, openness, and genuine relationships between issuing businesses and their customers.

Individuals and households could equally participate. Each person provides a service or commodity, whether it be their time, their energy, their supervision, or whatever else they do. Even a stay-at-home mother is a business; she is in the business of raising children. She provides a service (comfort, security, and nurturing) and employs people (herself, her partner, and perhaps a babysitter) to do so.

CURRENCY IS AN AGREEMENT

The use of money itself is an agreement on an exchange medium (without the agreement, it couldn't be used at all). If an Internet engineer and a mining company decide to do business, where the mining company hires the engineer to create a website, should they use a contract that a third person has created for them? Should they call Canadian Tire to help negotiate? No, that wouldn't make any sense. Yet this is the equivalent of using a device like the US dollar, the Canadian dollar, the Mexican peso, the Japanese yen, Russian ruble, or any of the over two hundred fiat currencies operating throughout the world. All currencies have terms of use, which are important to define, so that people can mentally assign them their necessary values (which, remember, is all that gives them value).

The terms of use for fiat currencies around the world are not necessarily advantageous, or streamlined, for the agreement that the web engineer and mining company are trying to make.

The web designer might think to himself, "Gosh, this mining company is very profitable, and seems like it's on the up. I would sure like access to some company stock in exchange for this job." If the miners had the ability to quickly and electronically create a currency that was redeemable for public company stock, they could accommodate this request easily, without worrying about spending something they don't have. Or, perhaps the web designer has been paying attention to the scarcity of copper, so he requests to be paid in a currency that could be redeemed for the mined copper ore, or the smelted metal (whichever the miners happen to provide as a final product). The web designer might request this currency, which the miners would be able to provide, under the assumption that the commodity, copper, could become more valuable soon, and therefore the web designer's pay for this one job would actually grow over time, since the currency he has would be more valuable.

With the ability to set up creative forms of payment like this, the businesses involved can maximize the value they both receive, as well as ensure a more personal relationship, that extends further than a simple transaction. Customized business dealings like this could become very fast, too—with currency systems based in computer servers and operated by smartphone/ desktop applications, managing customized deals would not be very difficult. A "personal" currency system like this would make more sense, and be less complicated than, handling the constant and complex borrowing and paying off of dollar-based loans that currently takes place in medium to large-scale businesses.

The digital realm is the only place these "agreements" could operate. In the past, creating highly individualized currencies would have taken too long and would not have been practical for everyday use (imagine if you wanted to change some terms of a currency's operation; in the past, you would have to go back and re-print it entirely). Yet in today's world, with the Internet and streamlined programming, it would be easy to make changes to an electronically-created digital currency.

Customized Currency

Why not? We already create custom agreements with neighbors, friends, and relatives without using money at all. Agreements that work for everyone are intrinsic to the human psyche—it surely allowed early humans to cooperate more effectively, which better equipped them for survival. It is clear that the same desire and utility for cooperation still exists today. We need to work together, in order for the amalgamation of businesses which we call "society" to prosper. This is how human life on Earth has always worked, and how, as far as we know, it will continue to function in the future.

Spaincoin, Isracoin, Auroracoin, and a smattering of other crypto-currencies have recently surfaced which are targeted specifically for populations of certain nations. As Spaincoin and Isracoin suggest, they are for Spain and Israel, respectively, while Auroracoin was created for Icelanders. These currencies are designed to boost local economies, while being tradable internationally. In this way, national alternate currencies are designed to be used as agreements for trade, to benefit a specific nationality. Yet if there was a software platform for it, anyone could create a currency.

So again, why not? In addition to providing more flexibility, having more currencies in circulation would likely strengthen each one. They provide leverage and exchangeability between each other, and prevent too much of one currency from dominating the market. With a system like this, anyone who provides anything could create a currency with a few keystrokes.

We have seen coins, credit, barter, and paper money, some with massive, complex institutions built around them, but what we haven't seen is changeable, personal, flexible, and digital money. It could use the lessons and utilities of everything that came before it; and that would be it; the thing we have been looking for to free our economies: the future of currency.

Part II: Modern Currency

"No State shall...coin Money; emit Bills of Credit; make any Thing but gold and silver Coin a Tender in Payment of Debts..."

Article I, Section 10, Clause 1 of the United States Constitution

CHAPTER SIX

A SHIFTING SYSTEM

Money itself will soon be as flexible as the software and instant communication of the Internet. New inventions like Litecoin, Dogecoin, and Bitcoin are harbingers of the money of the future.

THE FUTURE

The future of money, according to many Marxists, is none at all. In a vision of the Communist world to come, money, as well as ownership, is ultimately unnecessary and only spawns animosity between people. Yet 171 years later after Marx, a moneyless vision of the world has not come to fruition.

Money continues to be a necessary measure of value and therefore of wealth. It gives us the ability to procure the goods and services we need without having to barter. This is the result of specialized labor. Because our individual jobs and roles in society, as individuals we find ourselves with a surplus of certain types of labor or resources. Money, therefore, is the best tool we know of to bridge this gap of space and time.

Yet money became more than just a device for exchange; it became, in the past few centuries, a tool for confiscating wealth and consolidating economic power for governments, whose primary interest is defense. We find ourselves with a system of money that now makes little sense in terms of real value and mutual benefit for all (as it once did). Therefore the future of money is likely to return to a simpler, truly value-representative form, whilst simultaneously retaining the innovations and useful components of modern banking, such as electronic money creation.

MONEY ONCE HAD WORTH

The first paper money was "representative;" it was symbolic, and redeemable, for an actual commodity such as gold or silver. Yet precious metals proved to eventually be unwieldy for global currency trade, and largely unnecessary for personal transactors, who almost never redeemed their notes for gold—there was no reason to. People found the paper money to be far more convenient.

Money transformed into what we call "credit," which allowed "fractional reserve" banking to replace full-reserve banking. This allowed financial institutions to grow like weeds, and ultimately, allowed governments to control their domestic economies (in the sole interest of out-pacing other economies, so they could win wars).

There may have been a time when economic competition was a necessary measure for defense. But the time when regulation served the needs of the people, and was acceptable to the human consciousness, is over; just as the era of imperialism is over. In our time, economic regulation/centralization doesn't really make much sense (unless we agree that World War 3 is inevitable). Computer technology has exploded across the world and enabled idea sharing on an unprecedented scale, giving us a global cultural engine called the "Internet." This technology has not only made economic regulation more difficult, it shows how it's impractical and burdensome to a connected world; in fact, the Internet's main enemies are governments.

And solely because they control money, we have been unable to apply our incredible technology for sharing and mutual benefit to our system of money. Money continues to be tied up in banking, which is controlled and ultimately protected by government. The creators of actual wealth, businesses, have no part in the process of money creation. Their only allowed role in money's life cycle is as generators of debt, which guarantees that the arbiters of money (governments) receive their "share" of the wealth (since all debts are ultimately owed to them).

In reality, this system is built to ensure that every single unit of currency covertly returns to the institution that created it. And the only way it can continue to work is if the total wealth being generated consistently increases. Ergo, our system of money and banking relies on an unsustainable principle.

Yet shouldn't wealth, value, and credit should flow directly from usefulness, from practicality, from anything that serves other people? The people qualified to create currency should be those who provide goods or services. This does not exclude governments and their militaries, since they provide the service of protection for their peoples. Yet the ability to create currency need not be restricted to these entities; especially now that technology allows money to be created instantly, and transactions can take place over any Earthly distance.

CREDIT CLEARING

Many communities around the world make extensive use of Local Exchange Trading Systems (LETS), which allow people to locally trade goods and services without having to use money. These systems distribute imaginary "credits" to members and keep track of transactions. Using a LETS, no one has to use cash. The members get to keep their fiat money for purchases outside the LETS network, while inside the network, something extraordinary happens.

Because the LETS are transparent (anyone in the network can view any transaction), businesses and people naturally feel better about transacting with them, and it motivates people to help one another. People feel good about honesty—and when they see that their publicly-viewable "credits" increase based on the quality of their products or services, it naturally motivates them to do more business within their community (since it makes them look good to their peers). LETS have the added benefit of keeping people honest with each other, and quickly rooting out deceitful participants, who are quickly and easily spotted by the other members.

LETS are also interesting in that they often have very specific parameters for use. For example, some LETS have upper limits to how many credits one person/business can hold; if someone goes over their limit, they are obligated to spend some of it until they go back under.

Because a LETS does not require an outside or initial source of capital (money) to operate, it can economically bolster cash-poor communities by allowing them to extend their purchasing power with each other. These LETSystems act as "Credit Clearing" circles, which leave everyone "square" (with no outstanding debts) at the end of the day. Any other money that comes into the community can therefore contribute to the mutual wealth of the LETS members. The smart and simple innovations of LETSystems could be incorporated into future monetary models.

CRYPTO-CURRENCY

On the Dogecoin website, a video set to cheesy music depicts Shiba Inu dogs hitting what look like computer GPU units against rock walls and Dogecoins popping out. Dogecoins are not actually made this way, but the video explains the concept of digital creation: coins are "mined" into existence by computers that try to solve a complex encryption. When the computers finally crack a code, a certain number of Dogecoins are created. This is a simulation of the rarity of certain metals as used in commodity currencies of the past. This process, and variations thereof, is currently how all the major crypto-currencies are created.

Bitcoin, Litecoin, Dogecoin, and others are examples of electronic money being successfully made scarce and valuable. People use them because they are de-centralized. That is, no one party controls their manufacture and/or terms of use. They are also electronic, which means that transactions, and storage, are all done instantly through computers. The mutualism that crypto-currencies are designed to bring about is amiable, but there are also disadvantages to using them.

For one, accepting any crypto-currency is risky. Their values constantly fluctuate based on demand; a lack of concrete valuation or a redemption aspect leaves them vulnerable to wild swings in value. Another vulnerability is the fact that there is no buyer protection (since the currencies are totally de-centralized). Without a self-regulating system of value and personal guarantees for currency, electronic forms of money like Bitcoin tend not to become mainstream or widely-used currencies. They stay within a fringe niche without attracting users in significant numbers.

Crypto-currency has taught us important lessons about the modern monetary world, however. It shows that people are ready and willing to use something other than national currency for money. Innovative new tools for doing business and making purchases have made it into the public consciousness. Innovators are also willing, and passionate, to create new systems of money and credit on their own, using modern tools accessible to everyone. Finally, it shows that the Internet and communities created therein can operate independently of government control. Bitcoin, the first crypto-currency, created by Satoshi Nakamoto and a team of others, found a worldwide user base.

The more time that goes by, the more we get used to and comfortable with the Internet and its conveniences, the more people are ready for new monetary conventions—they just need to arrive in more accessible forms. A Bitcoin wallet is difficult to understand for the average person, as is the Bitcoin creation process.

We need simpler and easier to use electronic currency.

CHAPTER SEVEN

CURRENCY TECHNOLOGY

More and more, technology and biology are combining. Currency is following the same trend; it is due to be integrated into our very selves.

BEYOND INDUSTRIAL

"Technology" was not really a term in the English language until the 19th century. Things like trains, steam engines, factories, farm tools (think: the cotton gin), and more importantly, the bicycle, first appeared in that century. Before the inventions of the industrial revolution, technology had remained, at least for most people, very similar to that which had been used... well, forever. From this point, right around the introduction of the contemporary steam engine in 1781, things accelerated exponentially.

The bicycle in particular is important because it was one of the first examples of the technological trend toward miniaturization and simplification. When the modern bicycle first appeared, people were dependent on horses for commuting or other travel. Bicycles made it easier to get around, increased people's physical "range," and became inexpensive not long after their mainstream adoption. They were also small, and continued to get smaller.

Many of the principles that the bicycle embodied began to reach other industries; especially communications. Telegraphy had already been used for several decades at the time, and international telegraphic communication was common by the 1860s. This was a huge improvement over the "snail mail" system that preceded it, where sending messages was essentially dependent on couriers, and continued, as time went on, to become quicker, faster, and better. Telegraphs led, eventually, to telephones, and these led to fax machines. When the 20th century began, communication technology exploded with innovation, with everything from FM radio to cell phones eventually rising into contemporary use in little time.

Then, in the 1980s: the Internet arrived. The "Information Superhighway" combined visual, auditory, and text communication, and opened the doors for a whole new era of content sharing. It also redefined the meaning and practice of art, and opened up even more venues for connectivity; video games, social networks, and online collaborative tools, to name but a few.

Computers made leaps and bounds since their arrival in the 1940s, and along with communications technology like the Internet, exploded into commercial availability in the 1980s. Even then, advancements in transistors, conduction, power use, and storage all contributed to computers getting smaller and more powerful every year that passed. This led to the next big leap in consumer technology: the smartphone.

"Smartphone" is already a ubiquitous term, but describes a relative newcomer to modern life: a tiny portable computer that enables its user to connect to the whole sum of human knowledge and the nexus of all communication, the Internet, with just a few touches. According to Pew Research, 90% of American adults have a cell phone, while 58% have a smartphone, and that has been increasing every year since Apple introduced the iPhone in 2006.

In 2012, a new portable device became publicly popular. It was called Google Glass, and was a tiny computer worn on the head with a display directly in front of the eyes. It could do almost everything a smartphone could do, except it operated mostly by voice command rather than tactile input.

Because it was wearable, you never had to take it off or take it out of your pocket to use.

Technology is not only getting smaller in size, it is becoming integrated with our bodies. Organic light emitting diodes (OLED), made from organic materials, are the next big thing in display technology; not only are they more robust and can be used to make displays that fold or bend, they also have superior picture quality and use less power. We are discovering, milestone by milestone, that there is an inherent genius in nature, and technological developments are following her lead.

In fact, most trends in nutrition and health are moving toward "natural" and "organic" diets and "natural" remedies for illness. Many people are choosing to educate their children in a more "natural" and "organic" way, with amazing results. We are beginning to see the power of following our design; we are discovering how smoothly things run when they follow their natural shape and purpose. Despite its complexity, efficiency is built-in to nature. The more we study and learn about it, the clearer the genius of biology becomes.

Money's Personal Integration

Just as technology becomes physically closer and closer to our bodies and less complicated to use, money and the way we use it has followed the same trend. More mobile payment options, such as credit and debit cards, appeared around the 1980s, and these, along with banks themselves, became digitally integrated.

Banks realized long ago that people do not like to deal with complex money systems and accounts. They want simple forms of payment that are safe and fast—the widespread use of paper money in the first place occurred because people found it safer and simpler, as opposed to the complex and dangerous trades using gold and silver.

Credit and debit cards were the next step in this simplification, which made money even safer. Cash can still be stolen, and is lost forever if it is. Cards, however, less risky. One or two purchases could be made by a thief before the owner notices that the card has been taken. In the United States, most banks will reverse the purchases, and the bigger banks usually notify and cooperate with law enforcement to catch whomever made off with the card.

Along with safety, credit and debit cards are getting easier to use, too. Square opened up credit card acceptance to smaller businesses and individuals, and advances in Near Field Communications (NFC) and Radio-Frequency Identification (RFID) made it possible to transact without swiping a card.

Using these technologies, some smartphone apps allow bar and pub customers to start and maintain a tab without having to physically leave a card or other form of consideration at the bar. The same technology has been used to create applications for store credit.

While money converges with technology, technology converges with us. It makes sense that personal currency should become "integrated" into the technologies we use every day—since personal currency is a usable representation of the value which exists inherently in any person that provides something (i.e. a product or service) to society.

Money is not only due to become integrated with technology; it will be integrated with our lives, even more so than it already is. Each person is a business anyway, and value, which is inherent in us, in our means of production, should be spendable. This is why personal currency was not a possibility until now—until the arrival of not only the Internet, but of portable information delivery systems like phones and tablet computers.

The future of personal technology is unclear. Soon, computing devices and RFID readers might literally move inside us; we could have technology grafted into our flesh, and controlled by our own nerves. Or, instead of portable machines processing information individually (personal computers), will instead move into the "cloud," where centralized supercomputers and super-servers are instead accessed by individuals remotely, through individual connections rather than individual devices.

Either way, the prevailing technological trend is toward lighter, safer, and easier to use. Currency, like most other aspects of modern life, will likely adjoin the most popular technologies. Combined with transparency, like that used by Bitcoin, it would be centralized in a way that takes the technical burden off of individuals. It should, instead, empower them to make trades and transactions on their own terms; and to free the unique value that their business provides the economy.

CHAPTER EIGHT

A CURRENCY REVOLUTION

How this new concept, "Personal Currency," will change how we view and use money in a dramatic, exciting, and profitable way.

YOUR BUSINESS: YOU

John Locke, English philosopher, said: "Every man has a property in his own person. This nobody has a right to, but himself."
The American Dream is owning a business. Having a successful business is to possess success itself. What else is there? Since your business is you, the "property" inherent in each human being is their business, or their "means of production."

Income, or the reception of wealth, is directly correlated with the general perception of value. If people perceive your product or service to be valuable, they will respond by buying it, which is income for your business. Even working for some larger company is a transaction—that company, and the people comprising it, are buying your service or product; in this case, your time and energy are the products.

We all have a penchant for some kind of work; some calling on our lives that we feel drawn to. These inspire the genesis of small businesses all over the country, which in turn become larger businesses, bequeathing individuals with greater wealth, but more importantly, proportionally greater means of production.

Part of the American ethos has always been a belief in the natural right to freedom. If individuals have the natural right to freedom, and individuals are businesses, then businesses also have the natural right to such. This has always been at the heart of Libertarianism and Laissez-Faire ("let it be") economics, both of which have markedly influenced alternate currency movements. Alternate currencies enable businesses of every size (from individuals to corporations) to experience greater freedoms and to make more mutually beneficial transactions with each other.

Alternate currency is not a passing fad, nor is it a fringe practice without traction in mainstream culture. It is the result of an economic reality: the state-sponsored money we use is simply not the best tool for transacting—as a result, more people turn to alternate currency as a solution.

Alternate currency itself is a result of what famous economist Adam Smith called the "invisible hand." He believed that an economy self-regulates if allowed to do so—just as natural systems find equilibrium and mutually beneficial relationships (like the transactions between beehives and local plants), so do human systems—as they always have. People sell wherever wealth exists, and it flows from business to business based on the needs within the system. Alternate currencies, which are just different ways of expressing wealth, arose out of the need for more local cooperation, more freedom, more privacy, and less outside control.

A natural equilibrium forms wherever there is a diversity of resources. And personal currency would allow for equal diversity in the expression of these resources. This would allow for "fair" trade of all kinds to prevail throughout a personal currency economy. However, control, regulation, and centralization of any kind has always failed to produce successful long-term economies.

SCRIP

Money whose use and issuance is confined to a certain environment, such as within a company or other organization (like a military), is called "scrip." This type of money is easy to issue and ensures that people spend their wealth in a certain place or time; for example, remote mining towns in the 19th century would pay employees in a company scrip that forced people to buy company goods at inflated markups, thereby keeping supply costs lower and wealth within the company. A less pernicious example is military scrip issued to soldiers, designed to prevent money from reaching enemy forces within a local economy. Scrip was also used in Europe just after World War II, as a temporary measure to help civilians and military personnel alike rebuild the economy.

Some scrip would have a time limit for its use. During the Vietnam War, US soldiers were given scrip that could be spent in local Vietnamese businesses, but expired after a set time (usually a little longer than the leave period). This helped money stay both within safe communities and prevented it from falling into the hands of the Viet Cong, the guerrilla fighters opposing the US forces in South Vietnam.

Gift cards are modern examples of company scrip. While they are almost always tied to specific dollar values (in that you exchange dollars for the gift units at face value), they allow exchanges to be delayed. As of this writing, there is about $8 billion in unredeemed gift units in the United States alone.

In an electronic personal currency system, personal currencies would be able to act like scrip. Time, place, and people limits for their use would allow personal currencies to be as flexible as needed; whether for a single transaction or many.

PERSONAL STOCK

Shares of company equity, or stock, have become a cultural fixture, although a misunderstood one. They are actually a form of currency. When a person owns stock in a company, they own a piece of the company itself—anything the company makes, buys, or owns outright is owned by every stockholder. Sometimes, company executives and other employees are paid with stock shares, which can be very profitable in the long term—or disastrous, depending on how well the company as a whole operates.

In any business, somebody owns the shares of stock—in smaller businesses, this is often the owner or founder, whose personal assets are either liquidated into or tied together with the company. The advantage of taking a company "public" (allowing anyone to buy shares of the company stock) is that the company immediately makes a great deal of cash by selling itself.

Personal currency could easily replicate this, except with much more flexibility. A company of any size could "sell itself" ("go public" in corporate speak) for other currencies, if it wished, and therefore obtain more purchasing power in exchange for the confidence of others. This, just as it is for stockholders already, could be a very lucrative relationship for all involved. While there are certain restrictions on holding stock (such as "freeze periods" where no one can sell the stock), it is one of the most flexible value-transporting tools out there, and can be traded for other stock at values tied to the US dollar. Personal currency could allow people to trade stock at common exchange rates that have nothing to do with dollars—perhaps an oil company and its machinery supplier might price their respective company stocks in a currency they jointly create, in order to mutually strengthen their businesses every time someone makes an equity purchase of either one.

Even without stock options, businesses could create currencies within their supply networks and other partners. A bakery could create a joint currency with a coffee house across the street. The currency might have an attractive exchange rate from some other personal currency, such as the supermarket's—thereby encouraging people to buy it—and would therefore benefit anyone using it at the bakery or the coffee house (since the "prices" would be lower if you use their currency). It would encourage people to spend money at both. A win-win for everyone involved—even the supermarket, which benefits from having their own currency circulated.

A COMBINATION OF PRINCIPLES

These different forms of exchange are all forms of currency, and therefore, could be replicated by anyone creating their own currency. A personal currency network should accommodate all of these different forms of valuation. There are as many types of business as there are types of people—valuation should be just as flexible.

Personal currency would serve two purposes: one, to free businesses (people) from forced dependence on dollars. In our contemporary economy, to "start a business," (that is, to operate independently from anyone else), you have to either go into debt or rely on an employer for some time in order to amass enough wealth to begin doing business. With personal currency, however, there would be immediate access to real money for businesses. As a result, many would be able to start their businesses much faster and without the burden of immediate debt; thereby making more products and services available for everyone.

Much like how micro-businesses have flourished on the Internet, using tools like Amazon, eBay, and social media, digital personal currency would enable much faster and less hindered transactions. This would not only allow people to monetize their businesses faster, it would counteract the often entropic inflation that our economy tends to experience, since prices would be lower. Lower, yet stable prices help even more businesses to grow, since it reduces the cost of getting started.

And how would "lower prices" work in an economy with thousands, if not millions of personal currencies? Simple: it would mean that less of any consideration would be required to purchase anything else. Currencies would naturally find equilibrium with each other, just as commodities in a commodity-based system do. Indeed, you could think of the dollar as a commodity that everyone measures their goods and services against. It would work the same way in a system using personal currency, except each currency would be measured against every other, and since every currency is worth some amount of a service or goods, the "real value" of those goods and services would be what determines their purchasing power. This combines principles of barter (direct, quid-pro-quo trade), representative money (like gold-backed currency), and the IOU principles of fiat money.

A proposal for a personal currency network's basic guidelines:

1. **Consent:** no forced monetary use. Money is an agreement.

2. **Reputation:** a system by which users and creators of currencies give feedback resulting in a reputation score. Yelp is an example of a business-rating service that communities use very effectively to push their local businesses to offer better products and services. eBay is another example of an in-depth rating system that rewards honesty and quality.

3. **Valuation:** the ability to create currency directly representative of goods or services, and is redeemable for such. The ability to value personal currency in something more abstract—such as equity in a company or another currency. Flexibility is what makes money useful.

4. **Flexibility:** specific terms for each currency's use. An example might be currency which auto-redeems if certain criteria are met, or currency with an expiry date.

5. **Self-Regulating:** A common, open, and transparent system in which personal currency is created, stored, and transacted (like Bitcoin's blockchain), but is otherwise not interfered with at all.

With these five basics, an interconnected system of alternate, electronic, personal currency could grow and flourish on its own, and help businesses of individual and larger sizes grow faster, collaborate better, and trade with far greater flexibility than they otherwise could.

CHAPTER NINE

CAPITAL LIBERATION

Personal Currency would benefit businesses by liberating their capital, opening countless opportunities for trade/exchange, and transparency.

WHY ADOPT PERSONAL CURRENCY?

The reason currency has rules is to prevent abuse. It is an ordered system designed to prevent chaos, like all ordered systems. When they start to break down, or they are found to not serve everyone as effectively as intended, most systems are revised—in the case of modern money, though, the necessary revisions have not yet materialized. While different methods to value money have become mainstream throughout the years, its fundamentally centralized nature means that it comes from only one source—the Federal Reserve, which therefore, in a hidden, roundabout way, controls all of the wealth represented by US dollars.

It is the origin and method of creation of the US dollar that hobbles it. Since the dollar is designed to return to its source, its ultimate benefit is not for the businesses who use it. It must eventually leave the "custody" of anyone who has it. In fact, it is designed to; as an instrument of debt, every US dollar is owed to somebody (ultimately, again, back to the Federal Reserve which created it by lending it), and it will devalue over time anyway. Inflation is a silent, hidden tax, which is artificially perpetuated by the Federal Reserve (as they create more money, the money already in existence loses more value).

Of course, if the amount of goods and services available in the economy continue to increase at the same rate as the money being created, then inflation will remain at zero. But our economy has not followed this pattern. Instead, the economy grows more slowly each year, while the Federal Reserve continues to print money by buying more bonds.

This makes businesses subject to the will of the US Government. Controlling the direction and flow of money enables the "Fed" to exercise some creative control over the economy, yes, but this also creates an artificial environment for businesses.

Over time, the US government cleverly created a monetary system that resembles a large reservoir that flows out to rivers, which conveniently meander their way back to the same reservoir.

This is why enterprises in the United States do not grow in the organic fashion that they once did, or that businesses elsewhere do. There are a myriad of regulatory laws controlling taxes, international trade, and production, true, and these do affect the growth of businesses. But the actual nature of our money is another matter entirely. Every dollar created is like an iron filing creeping its way to a giant magnetic rod: the federal government that created it.

It's simply built-in—every dollar created is owed back to the Federal Reserve eventually, since they only lend them out; they aren't given away. Therefore the dollar is a poor store of wealth.

This is why a personal currency system would be so valuable to modern businesses. A new, truly modern monetary system that allows people to simply and easily build themselves up, and not have to go into debt to do so, is desperately needed for all. Personal currency would be just the vehicle to help business not only store wealth with greater security, but to foster greater cooperation with partners, customers, and everyone else who uses their currency.

BUILDING A CURRENCY

Part of the dynamic of personal currencies would be "reputation building." Just as a business builds a reputation on Yelp or with the Better Business Bureau (BBB), which strengthens their brand and encourages people to buy from them, currencies will need to be trusted in order to work (the only reason that mainstream currencies like the dollar work is because people trust them in the first place).

Just as a brand must be tended to and grown in the consciousness of consumers, personal currency, which would serve as a brand's representation of value, would need to be "grown." A currency's reputation is actually what gives it its power to be used as money—which is why building a good reputation is essential in a personal currency system.

Building a reputation is essential in business anyway. Wouldn't it be ideal, for both sellers and buyers, that a good reputation should directly correspond to more wealth? Equally, it is logical that a poor reputation should negatively affect wealth. Businesses are only as good as what they provide for people; their services.

"Service," which takes the form of both physical products and of amenities that businesses provide, is the true source of wealth and value. We attribute the most value to things which serve us the best. Whichever need those things meets determines how valuable they are (which is why diamonds, which serve no practical purpose but serve to solidify extremely significant relationships, are priced far higher than water, which is essential for basic living—relationships are a "higher" need).

This is why becoming "mainstream" shouldn't be a bad thing. Business growth allows people to be better served; better facilities, more efficient delivery, more personnel, and better technology should only help businesses deliver better results.

Another benefit to having larger organizations participate in personal currency will be the ability to have stronger currency. Businesses composed of groups of people can participate in more, and larger, transactions, which builds trust faster. For example, an individual selling coffee machines door-to-door can only participate in one transaction at a time—therefore building trust only one person at a time. Yet a company which employs several salespeople who all sell door-to-door at the same time will build trust several transactions at a time.

In order to secure loans, phone plans, and other trust-based commodities in the traditional economy, we already use credit scores. Credit ratings made a huge impact on how business was done in the 20th century and beyond, and helped determine who had access to what. "Reputation" is a more direct form of "credit rating"—and instead of coming from an agency and from certain high-profile "reporting" companies, it comes from marketplace peers; the people that you choose do business with.

Individuals are certainly able to build their own currencies' reputations, and in some cases, standing on your own may be more advantageous. Individual currencies will certainly be faster to create and redeem, since companies/corporations will have to manage more complex currency systems and larger amounts.

Yet in either case, because personal currency will allow reputation to factor directly into the value of a business's goods, services, and credit, building reputation will be more important and more effective than ever before.

MORE TRADE OPPORTUNITIES

Because personal currency will free the credit of both individuals and businesses, more exchanges can take place. The implications of this are huge—the reason that the government tries to encourage spending during economic downturns is because it really does stimulate the economy. More circulation means that more goods can be produced, more services can be sold, and therefore, more jobs will be available (therefore, more spenders enter the market—and even more value enters the system). Imagine if the US dollars requirement were lifted from spending, and suddenly, a robust economy on all levels (local and national), is possible.

Essentially, everyone has more money, since they are free to use their "credit" without possessing dollars. More money (more value in the market), means that the economy grows and moves, which is good for everyone involved.

Personal currencies allow businesses to adapt to changing circumstances faster as well. Sometimes, wild market changes, like the release of a new and extremely innovative product, can disrupt an entire industry. For example, a new computer technology developed by one company that completely surpasses all competition might render competitor companies' currencies less valuable for awhile while they adapt (much like how stocks go up or down in value based on a business's efficacy).

Yet a company with diverse currency holdings can easily survive such a scenario (and others). Let's say that company A has a product or service that becomes wildly successful, while companies B and C (A's competitors), are left scrambling to come up with a solution. Another business with huge cash holdings in company B might suddenly find their stored value diminished as company B is out-sold by A.

Yet if this business has holdings of currency in more than just B, in A, in C, D, E, and F, and perhaps in other industries, they find themselves "insulated" from such a situation. While the values of B and C's currencies may diminish as public interest in those companies drops, and possessing currencies from unrelated companies makes the loss in value less significant—plus, if some currency from company A is in the business's "coffers," they might actually profit from the event—since the currency of A becomes more valuable even as demand for its competitors' diminishes.

Or, if by some chance a huge amount of company currency were all redeemed at once, a company holding reserves of other currencies could simply exchange the equivalent amounts for their own currency, returning the purchasing power to the currency's originator.

Does this sound familiar to you? If you've ever traded company stock, you know that it is very similar. The stock market is a type of currency exchange where "cash" is traded for shares in a company, in the hope that those shares will increase in value over time (and therefore provide an opportunity to profit on the initial investment).

A personal currency is like a direct "share" in a company's interest. Rather than having to sell these shares for "money," in order to buy other goods and services (which would make them indirect shares of a business's value), personal currencies are money, and can be used as such to purchase other currencies or to purchase goods and services. This simply removes the traditional stock-trading middleman: a government-created currency.

Combining stock and currency allows a company's credit to strengthen without the company doing anything. As it circulates, it gains reputation naturally. An example is before the 20th century, when new currencies would appear alongside old ones, due to a certain new need or popularity. A real example would be the various East India Companies, who all produced their own currency. These currencies, originally designed as a local colonial currency, steadily grew in usage around Europe.

Because European trade with Asian merchants became more economically important through the years, East India Companies' self-minted money grew in demand (and esteem), giving it far greater value. Part of what allowed these companies to last so long (the British East India company lasted 274 years) was that they were able to not only create and hold their own money, but they actively and earnestly invested in each other's.

Money diversity in India was not a disadvantage to buying goods, either; the coins and notes of different monarchies, governments, and companies were welcomed when trading, as merchants understood that possessing a diversity of money was good for business. At that time, it would have likely been disadvantageous to hold only one type of money. If the company or government that produced that money suddenly lost its reputation, was destroyed, or became otherwise untrusted, your stores of wealth would similarly become defunct overnight. Just as in modern stock markets, it is less risky to hold equity in a diverse range of businesses.

It is the difference between holding the wealth of a single business, which is just one cog in the massive machine, and holding the wealth of an entire economy. Indeed, when everyone is invested in each other, it becomes less likely that any one might fail... which benefits everyone.

The more pillars holding up a weight, the less likely the weight is to fall. Diversity of wealth, brought on by personal currency, can be one of the greatest advantages for any business.

Part III: The New Money

"The refusal of King George to allow the colonies to operate an honest money system, which freed the ordinary man from clutches of the money manipulators was probably the prime cause of the revolution."

Benjamin Franklin

CHAPTER TEN

YOUR CURRENCY

**You already have a currency. Every business
and individual has a currency; something their
business provides society.**

UNDERSTANDING PAST FAILURES

Despite the pervasiveness of money like the US dollar, there were recent efforts to create multiple-currency systems. Many alternative currencies cropped up again after the US Civil War and the World Wars, especially in the late 20th century—only to fail, or at least remain largely irrelevant. Even the most enduring, like Ithaca HOURs, failed to make a large-scale impact on even their respective local economies, which could still not survive without dollars.

But why?

In a word: *centralization*. Economic planning on the government level creates an environment which is incompatible with competitive currencies. It is not by accident, either; governments purposely insulate their economies from outside currencies so that their own economic assets are more stable and predictable.

This has been attempted since the invention of governments—even the Romans tried to do it. Historically, it failed every single time. Let us repeat that: economic intervention has never, ever worked in the long term. In fact, societies which practiced economic freedom and laissez-faire policies were economically successful, and would adapt more quickly to new conditions. Human economies grew and flourished for thousands of years before anything resembling modern governments appeared.

Yet as the world's population increases and economies bloom, stagnate, or shrink, ruling authorities only attempt stronger interventions.

Alternate currencies have simply failed because they are almost always valued in terms of the national currency, an inextricable instrument of debt, making the alternate currencies just confusing asides to the government-sanctioned unit of account. These not-dollars operate the same as dollars, but call themselves something different. They are a hassle to get a hold of (you usually have to buy them... using dollars), they are a hassle to manage, and adopters never invest much into them, because at any moment, people might stop accepting the alternate currency, rendering the investment worthless.

For an alternate currency to stand on its own, it should be priced in something other than the national currency.
It's because the idea of value itself is changing.

A NEW (OLD) IDEA OF VALUE

Money is so important to individual lives because people treat it as if it is wealth. Money and value are synonymous in Western culture.

It can't be stressed enough: this is a very recent idea. A diverse community of businesses using a uniform, standardized currency has never even been tried before the 20th century. Ever. Not even the Roman Empire, whose "denarius" widely circulated, used this money uniformly, all the time, for everything. Over short distances, such as within cities, the denarius was used with great success, almost as a community currency, keeping wealth within the locale. However, many other kingdoms and principalities, with little access to Imperial gold, would mint their own money from local metal sources or stockpiled wealth. These different forms of coin were used alongside each other. Over larger distances, people directly traded, using trade caravans to haul goods across the world. Such methods of trade were "mainstream" until the 20th century.

Indeed, by contrast, the economies of the modern world look streamlined and artificial. "Wealth" is no longer equated with utility, or any kind of directly accessible value. Wealth is equated, in the minds of Westerners, with cash, or with digital bank units stored inside computers. Money exists in numbers, quantities, which are very carefully defined.

Personal currency would serve to, among other things, bring an older idea of value back into the cultural consciousness—while combining it with modern monetary innovations and technology. Traditionally, humans considered whatever was useful to be valuable. Roughly, value=utility. If you remember the example of "diamonds over water" from earlier chapters, however, you know that for humans, utility is in a complex hierarchy. When our basic needs are met, such as sustenance and safety, we crave other things: love, friendship, self-esteem, respect. At the highest level, we crave creativity, morality, and the fulfillment of a mission.

We price accordingly. People with most of their needs fulfilled will crave the highest need, the one for spiritual, existential fulfillment (what Abraham Maslow called "self-actualization"). This leads them to pay incredible prices. Accomplished businesspeople will donate millions, in some case billions, to charity. Some people give up everything to become monks or nuns. More still will go to great lengths and down harrowing paths of spiritual enlightenment to find that something which they do not have, and cannot see, but desire "more than life itself."

Next down the list is Esteem. We donate most of our lives, our time, to becoming educated, then working in a job we feel suited to, or that we desire specific achievements in. We pay enormous prices for college, for training, and some sign on to military service, a massive sacrifice, in order to achieve what they want. Other people feel that their physical selves are what they want respect for, so they pay with hours upon hours of sweat and physical hunger to get the body they desire.

Then there is Love and Belonging. We will "do anything" for our families, and many go to extraordinary lengths to protect and serve them. Mothers and fathers alike donate countless hours and much of their earnings to raising their children, to meet their own needs for familial belonging, for intimacy, and for friendship. In this category also are the things that everyone does for their friends, people from whom everyone seeks approval and respect, and people go to great lengths sometimes to get it. Love and belonging are why diamonds cost so much more than water. Diamond rings are instruments of social change—they are used to "purchase" marriage. Because marriage satisfies a "higher" need, people naturally place a greater psychological price on diamonds.

Finally there are the basic physiological needs, like for water, food, sleep, safety, and general health. These things do not cost very much, although they are necessary for the most basic existence. As soon as these needs are met, humans inevitably begin to strive for love and belonging if they don't already have them.

Value is therefore a function of how rare and difficult something is to obtain and how much utility it has for our lives. And utility is simply the sum of how completely a service or commodity fulfills a need and which level of need it fulfills. For example, if there was a pill that completely and utterly fulfilled the need for a mission, purpose, and creative significance in life, yet was rare or difficult to create, it would likely be the most expensive thing in the world. Alternately, if a pill were invented which met your entire lifetime need for sustenance and sleep, it might be expensive, yes, but not nearly as expensive as the other pill.

This is because humans innately know that simply surviving is not living. Our higher needs deserve more of our attention, our time, our labor, and our wealth. That's what makes them valuable! And these things take a different form for each person. One person might find a career in engineering to be incredibly valuable and worth great sacrifice, while another finds a career in writing. Value is subjective to each individual, yet still operates according to the factors listed above.

People buy extremely expensive super-cars not to drive around, but to meet their need for self-esteem.

They pay for expensive, dangerous, and painful surgery to appear attractive.

They go to marriage counselors to save their most important relationships; to preserve families.

Still more is paid to gurus, teachers, retreat centers, churches, mosques and synagogues in hopes of finding meaning.

We will pay with our very lives in service of what gives us meaning. Because having meaning, to us, is truly living—and that's why money and value are actually not synonymous, and could never be. It's also why personal currencies, thousands of them, millions even, could exist together, simultaneously, exchanging freely, both for goods and for each other.

Two men might trade a cow for an ox, and agree it was a fair trade. Yet to one man, a cow is more valuable than an ox, and to the other, an ox more valuable than a cow. Value is subjective, and is highly dependent on where an individual's "hierarchy of needs" stands, which is dependent on countless factors within their own life. This is the very foundation of "economy" —the management of resources—and is totally organic. Meddling with it disrupts a highly sensitive natural order. Personal currencies would simply allow the exchanges that we all need to take place to happen faster. The men trading the cow and the ox are trying to accomplish two different things, yet are helping each other do so. One needs his field plowed, and the other needs milk to feed his family. The cow and the ox are just the tools they have found for these needs—and, finding each other, having different but compatible desires (one has what the other wants, and vice-versa), they are able to trade.

Subjective ideas of value are what actually allowed the two men to trade, and the same principle applies to personal currency. Many businesses may find "oil pesos" representing an oil company's interests to be incredibly valuable, while other businesses may not find much need to redeem them at all. Yet the interest and subjective value placed on the oil company's "pesos" by the first group would make them just as valuable to the second.

Therefore, the second group, while not needing any oil, would still feel safe and confident investing in or trading the oil company's currency. No dollars are needed to "price" personal currencies in. No "pegging" is required—only a simple rating system.

YOUR VALUE, YOUR CURRENCY

As mentioned above, the biggest obstacle to trading without dollars is seeing the value of things without dollars—to return to a more natural way of determining value: personally and subjectively.

It's easiest to do this by looking at yourself.

What do you produce? What is your business? This is how you can find the currency you already have. Creating a personal currency is a process of monetizing the value that's already present in your business (which is you). It might be an easy question for you… or it might not. If you own a lumber company, it might seem simple. Your company directly produces lumber. Your company can create a currency valued in quantities of such. If you are a hairdresser, perhaps your currency could be valued in quantities of haircuts.

What if you're a doctor? A counselor? A consultant? A teacher? None of these produce a commodity, they instead provide a service. Instead, their currencies can be easily valued in time. A professional's time can be very valuable, especially if they are particularly skilled.

There are uncountable ways to value currency. Many companies already create their own currencies and are highly creative in how they value them; they just don't call them currencies. Starbucks, for example, uses "stars," which are points that customers accumulate for buying their products on registered cards (free plastic cards with magnetic stripes, similar to credit cards). As customers accumulate "stars," they can spend them to purchase various store products. Most people think that the things they get for "stars" are free—but they aren't. Starbucks has simply created a currency that is priced in transactions, rather than dollars directly, and is a form of delayed value, since each star costs a dozen or so transactions. Then, any drink (and some food items) on their menu can be purchased using a star.

Starbucks has trained customers to think of the stars as "rewards," but they aren't. They are simply another form of currency, just as all loyalty programs are. Another example is airline miles. If this is unfamiliar, these are points accumulated for using air travel, designed to encourage the consistent use of whichever airline distributes them. These are used to purchase different things from the airline, such as additional airfare, accommodation, rental cars, meals, or in some cases, entire vacation packages. Some airlines have expanded their programs to include the ability to purchase a huge variety of products, including magazine subscriptions, music, movies, and countless other items.

Because these various programs originate and are used within their respective businesses, they get to make their own rules. United Airlines has an exhaustive list of rules (nearly 4,000 words as of this writing) on how their airline miles work. That's exactly what a personal currency is. A quantified, subjective value system.

Imagine what makes your business unique. What is the value you provide people? What is the "thing" they come away with after doing business with you? This is the key to finding your currency. How do you serve? What is the service of your life, your business, your company, your time? You already have a currency. In fact, you might have several... they just haven't been "monetized" yet.

Now imagine that *every* business fleshed out their personal currency. It would not only mean a change in the way we handle money; this would usher in a new type of economic system.

CHAPTER ELEVEN

A NEW ECONOMY

Personal currency really represents a new kind of economy—a new way of doing business that just makes more sense.

A MATTER OF FLOW

In the centralized monetary system, where money is value, wealth flows from one source (the central bank) and must trickle downward, flowing fastest when it is being lent and therefore becoming debt. In this system, because more money owed means more can be created, consumption is highly encouraged, and is seen as the key force for economic growth.

This is illustrated by a pyramidal shape:

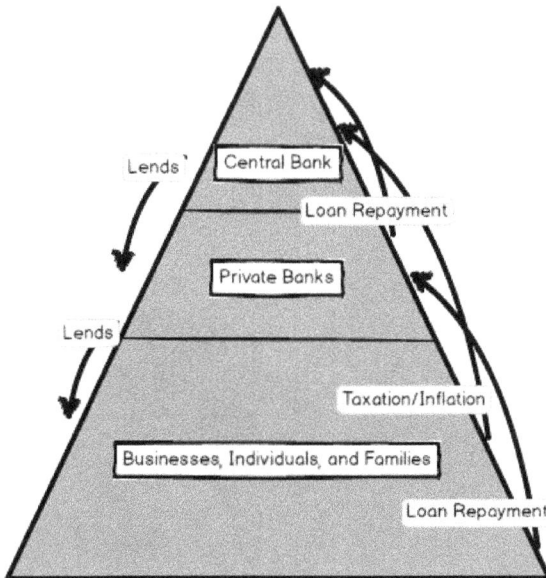

Federal Reserve System

A system of personal currency, one where value originates in people themselves (and is worked, or expressed, through business), would flow in the opposite direction. Each person would be like a bank that creates their own currency, redeemable for whatever they provide the economy (or simply created to express value that two parties would not have otherwise been able to exchange). When value is quantified by value providers (businesses) directly, it is limitless. Businesses will continue to produce their product or service for as long as they live. This means that a stream of non-debt based legitimate currency can flow out of them continuously, profiting them, and growing the wider economy. Everyone benefits, as more money enters the marketplace all the time. Production, rather than consumption is the essential power of this economy.

It would take an inverted pyramidal shape:

Personal Currency System

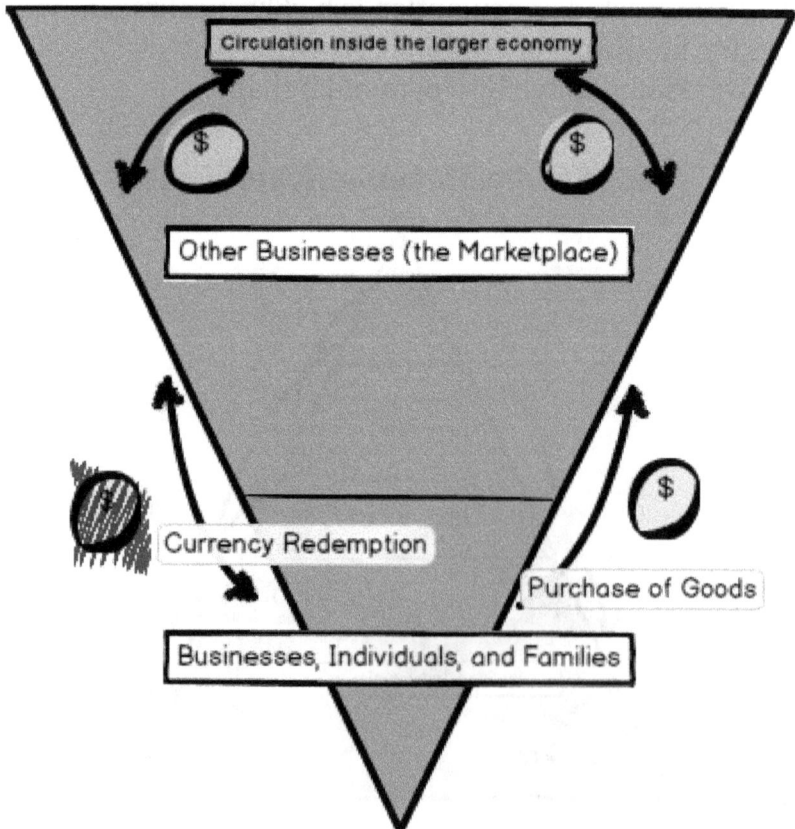

Circulation inside the larger economy

Other Businesses (the Marketplace)

Currency Redemption

Purchase of Goods

Businesses, Individuals, and Families

Currency inside this type of system would move cyclically, from creation to redemption. This generates a cycle of goods and services that travel outward from each currency, both to other businesses and to the currency's original creator. Conversely, in a centralized currency system, goods and services can only proliferate relative to the total money supply.

The argument here is that when the economy is treated like a closed system, a repetitive mechanical machine, it is rife with inefficiency. Remember, the purpose of economy is to effectively manage the resources of society. When resources are misappropriated, and flow in unnatural patterns, there's something wrong. In this case, the fiat currency that modern economy "travels" on is flawed and restricted, and has reached its useful limit.

Manipulating monetary policy, with the intention of creating a predictable, stable, and prosperous economic equilibrium, has failed. It hasn't just failed once, and recently, either: it has failed again and again, multiple times throughout human history. We have little excuse not to attempt alternatives to such.

Every year that goes by, it becomes clearer that the modern world is transforming into something wholly uncontrollable and unmistakably organic. An economy is still a system—albeit an organic one. It should be treated accordingly. Plants do not grow by yanking on them or commanding them to be or go a certain way. They grow when the proper conditions arise.

The "cogs" that connect together to form the whole economy are made of businesses, which are made of people. People themselves are complex systems of biology, emotions, and thoughts; they are organic, not mechanical, expressions of energy. Yet the systems we created have reflected the apparent belief that people are mechanical and repeatable, like wind-up toys.

Organic systems promote fecundity. A feature of any organic system is rapid growth that adapts to changing conditions. A more "organic" currency system would do the same—without restrictions to growth, or "dams" marring its flow. This was how Bitcoin gained so much value later in its life; its decentralized nature meant that its worldwide, diverse user base could expand without regulatory roadblocks. In contrast to Bitcoin, however, diverse personal currencies with multitudes of uses and valuations could expand, change, and grow beyond any modern or historical example.

PAY THE PEOPLE

Creating company currency means monetizing the wealth already present, yes, but it also includes the ability to build loyalty among people in the company. As stated in Chapter 9, personal currencies would liberate the credit already present in the unsold products and services that are otherwise just sitting around. This "potential energy" is often wasted, but could be going somewhere useful—whether to new processes, new equipment, or to plug gaps in infrastructure.

It can also go to the people working at the company. Using company currency strengthens it—the more of it circulating, the faster it will build trust, and the more momentum it gains. This makes all the currency that the company subsequently creates much stronger and more useful.

Companies might even use personal currency as a form of timekeeping. Imagine if a company that, upon hiring an employee, pays them in company currency equivalent to an amount of the employee's personal currency. Then, as the employee does work for the company, the company redeems the employee's currency for the work completed.

If the employee has to work more, he sells more of his personal currency to the company (which instantly "pays" him with company money), and the company redeems the extra "labor" that the employee gave them.

This instantly looks more like a working relationship between two businesses on equal footing, as opposed to the servant/master transactions that we often end up fighting against in the modern economy (and have been struggling with for centuries). This also enables a transparent exchange of labor for value, clearly showing what each party is giving and receiving.

Automatically, the "air is cleared" for everyone. No matter how large a company gets, everyone stays clear on what values are being exchanged—promoting honesty for all.

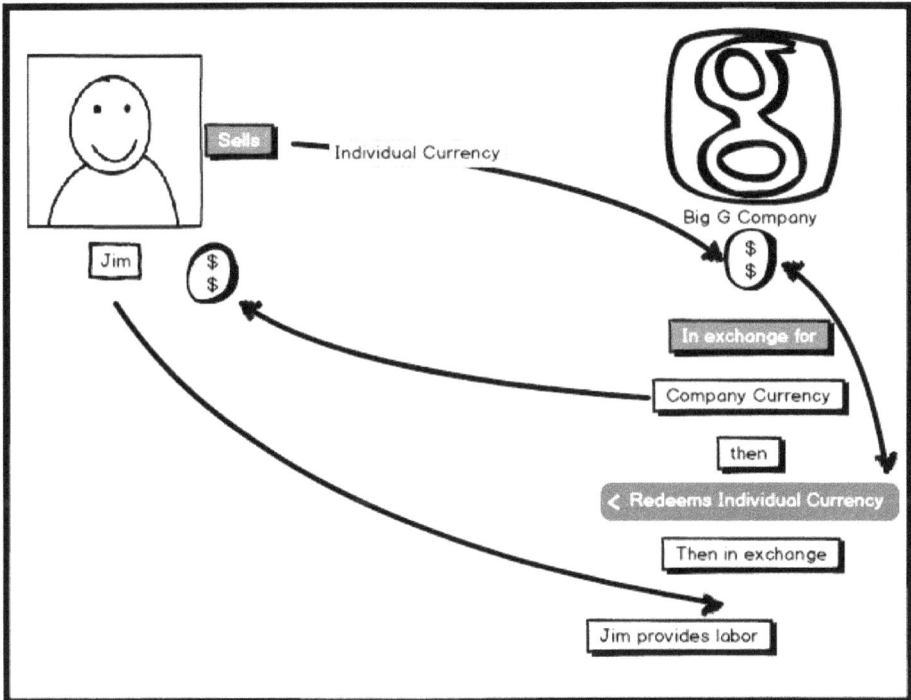

Rather than servants and lords, companies and their workers become more like true partnerships, where independent producers work together for a common mission in a common space, and as a result, produce far more than they could have done individually (while sharing the benefits in a transparent way).

Much like the human body, where each part has a unique and essential role that receives the resources it needs to function, the various people inside these "corporations" (corporate actually means "body") would contribute their individual values (their currencies) to the company, and the company (the larger body) would give them shares of its corporate combined value in return.

This relationship looks natural—like an energy cycle—indeed, what are money and value except energy? When we exchange goods, services, or tokens that represent such, we really just exchange different forms of energy. In keeping with the "body" analogy, currency serves the same purpose as blood. Blood carries nutrients and oxygen from the respiratory and digestive systems to other organs, where other materials, which in turn are essential to the lungs and digestion, are then deposited into the blood and transported back.

This process is intricate, with so many parts and roles to play, yet blood can support all of it. It is the medium that transports every form of essential nutrient to its proper place.

This is the exact role of money, and until now, it has not performed well. In the modern "body," there is an odd coalescence of "blood" in certain places and an almost total lack of it in other areas, which are no less essential. It is clear that a change is needed; a better form of "circulation" is required. When the resources of our economy (the "nutrients" in our "body") move the way they need to, it might finally flourish and grow, like a vine given the right conditions for growth, in a way that each part, no matter its size or function, might have enough.

Because it's not the economy that stands to gain or lose, live or die in the struggle for energy. "Economy" is just a human idea. The real, important, living factors that comprise this idea are businesses. Each person, every single individual, is one of those businesses. So in truth, when we discuss the need for economic change, "markets," "rates," and "fiscal systems" are not what we are discussing—we are talking about the lives of people.

CHAPTER TWELVE

A CURRENCY COMMUNITY

To adopt personal currency, a community of businesses—both companies and individuals— would have to participate.

BORN OF NEED

During the Great Depression in the United States, the US dollar catastrophically deflated (increased in value and in scarcity at the same time) as the demand for goods went down and the Federal Reserve was unable to artificially inflate the money supply (due to a strict gold standard being in place). This left many individuals and countless businesses severely lacking in cash.

In response, companies and local governments did the only thing that made sense at the time: they created their own currencies, which they called Scrip. Because so many companies were cash-poor, they had to either pay their employees in scrip or go out of business entirely. Being that there were not many sources of money outside these companies (dollar-paying jobs were scarce nearly everywhere), people accepted the scrip as payment rather than have no work at all. It worked well—better than expected. Because of widespread bank failures, people did not trust them as issuers of credit; many who had loans or stocks managed at banks were left with nothing to show when many institutions became insolvent. Scrip gave people the ability to purchase, both at individual companies in local areas and with each other. Scrip itself became a commodity that was highly valuable in areas without much access to US dollars.

In Germany and Austria, both during the same period and earlier, local authorities issued notes called "Notgeld," which means emergency money.

This money became the most stable currency in the region, as the German Mark plummeted in value and became practically worthless in the early 1920s.

While a similar desperate situation could happen in the modern world again, it doesn't have to in order for businesses to take advantage of personal currencies—people, especially companies, can begin creating their own currency right away, without having to immediately migrate away from US dollars.

AN EMERGENT COMMUNITY

As with any currency, a Personal Currency system will require an initial group of adopters. Currency only works when it circulates, and it will need to prove itself worthy of replacing the national currency for those who choose to use it. Not that it must replace the national currency—personal currency could be used alongside government money easily (some, especially those just starting to use alternate currencies, may choose to price their newly created currency in US dollars). It is more likely to coexist with government money in the beginning and grow into a more widely adopted exchange medium as time goes forward.

There are several ways that this community might form. The first is local community adoption—how alternate currencies have traditionally begun. This is where a group of businesses and individuals in one locale get together and create a local monetary system that operates alongside the national currency. These are usually meant to keep dollars in the community, specifically, and so are hardly alternative forms of money. Still, this could work well for personal currency. The advantage to trading with your neighbors is that you know them—trust is easier, and therefore everyone's personal currencies could build reputation and value much faster.

Another way that a personal currency exchange system could form is online. Bitcoin, Dogecoin, Litecoin, and others prove that a working wholly electronic currency works for the exchange of real goods and services. While the technologies are new and many are not user-friendly, the Internet is promising territory for currency, since it allows transactions to take place instantly and over any distance.

A community online could start by trading Internet-based work like programming, web design, art, writing, etc.

The most likely way a successful personal currency network would work is a mixture of these two. Auroracoin is an example of an electronic currency designed specifically for one area—Iceland—with the bold premise of first supplementing, then replacing the devaluing Icelandic krona, and eventually trading for other currencies. Another electronic currency that works both locally and across great distances is the Life Dollar, which originated in Washington State. Life dollars are managed by an organization called Life Currency Cooperative Exchange, which localizes independent community money while allowing different communities to trade with one another over distances. With a currency like Life Dollars, people living in the same locale can use currency face-to-face, then connect to others across the country whom perhaps haven't met but are all in agreement with the program's rules and principles.

Personal currency, as explained in Chapter 8, would operate under a set of principles; rules, if you will, that prevent anyone exploiting the system. These are, again:

1. **Consent:** no forced monetary use. Money is an agreement.

2. **Reputation:** a system by which users and creators of currencies give feedback resulting in a reputation score. Yelp is an example of a business-rating service that communities use very effectively to push their local businesses to offer better products and services. eBay is another example of an in-depth rating system that rewards honesty and quality.

3. **Valuation:** the ability to create currency directly representative of goods or services, and is redeemable for such. The ability to value personal currency in something more abstract—such as equity in a company or another currency.

4. **Flexibility:** Specific terms for each currency's use. An example might be currency which auto-redeems if certain criteria are met, or currency with an expiry date. Flexibility is what makes money useful.

5. **Self-Regulating:** A common, open, and transparent system in which personal currency is created, stored, and transacted (like Bitcoin's blockchain), but is otherwise not interfered with at all.

Devising and adopting a personal currency network is as much a function of morals as it is of practicality. The modern world is rife with deception, scams, lies, and clever folk who seek to, at best, take advantage of harder workers, and at worst, outright steal everything they can. Ruling authorities are no exception, and are perhaps the best example of why change is needed. Organizations designed only to fight wars and "regulate" trade (as discussed in the first four chapters), are, at the very least, worthy of suspicion and careful scrutiny.

Indeed, their resistance to meaningful change is an indicator that businesses themselves must find and act on economic solutions.

Law, as a way of regulating human behavior, is an expensive and burdensome method (try hiring a lawyer and this will become clear). Morals are one of the strongest motivators of behavior, and hold entire organizations together through difficult circumstances. When it comes to exchanging value (currency), morals are especially important, because having enough value is a matter of life and death, not only of convenience.

The good news, great news even, is that a system built inherently on trust is self-regulating. Dishonest businesses will not be able to operate in a personal currency network for very long, as the market will quickly lose confidence in their currencies, rendering them valueless, and the suspicion surrounding them will prevent people from transacting with them (this is exactly what happens to many businesses who gain poor reputations with rating agencies such as the Better Business Bureau).

Because transparency is built-in to a personal currency network, deceptions will be difficult and more easily found. Yet in order for this network to arise in the first place, a group of businesses with the desire to see operate in such a space is required.

An honest, hardworking, visionary, and bold group of businesses is needed to start—but as the institutions which hold up the old order slowly fade and break down, more people, searching for solutions, will seek out a better business space, and arrive "on the doorstep," so to speak. Everyone will benefit if a working, circulating system of new currencies already exists. It could grow exponentially, much like Facebook did when it became publicly clear that it was the most accessible and user-friendly social network in existence at that time.

Better to build new currency networks now, before bread and gasoline creep their way out of the average consumer's price range.

CHAPTER THIRTEEN

INDIVIDUAL USE

In a not-so-distant future, an example of one person using personal currency in their everyday life and business.

A COMMON PROBLEM

Sherri is a hairdresser who lives in a small, quiet community in Southern California. Rather than having a shop location, she carries tools with her and makes house calls—and has a high number of regular clients. Sherri's family-friendly and personalized service is beloved by her community, and she's often known as "the best hairdresser in town." Because of the distances between residences and commercial properties, the fact that she travels to clients makes her highly convenient.

Before personal currency, Sherri had a few business problems. To begin, her clients could book appointments, but sometimes would have to cancel that day due to circumstances, resulting in a wasted time-slot for Sherri and no money in return. Sometimes she would even be on her way to a client when they called to tell her they needed to reschedule. This was frustrating, and caused Sherri to lose a lot of time and money.

When Sherri began using a personal currency application to create her own, things changed. Using the app, she was able to create a clever work-around that benefited both her and her clients.

She set up her currency, gave it a name (TrimPoints) and some basic terms (redeemable for Sherri's haircuts, coloring, shampoo products, and hair extensions). Most importantly, she installed a new policy: now she should be paid upon booking, but now she will accept personal currency, as well as dollars if her clients want. In the case of a cancellation for any reason,

Sherri simply refunds her customer in an equivalent amount of TrimPoints. This function is dual: on one hand, her TrimPoints are directly redeemable for haircuts and other services. On the other hand, she still has a form of usable currency in-hand in exchange for her work.

When the client is ready to reschedule, they just have to redeem Sherri's points for the haircut/color/products/extensions they originally wanted. Anything extra is purchasable with either their own currency or any other they have accumulated. If the client decides they don't want to use Sherri's services, they can still spend her currency on something else! Especially in town, where the quality of her business is well-known; her currency is highly trusted as a result.

EVEN WHILE AWAY

Sherri, while very happy at home, decided to get away for awhile and travel. She'd always wanted to see South America, so she took three weeks off of her schedule and booked a flight using Walmart gift card credits, which she had bought using a local lumber company's currency (because Walmart, ever-resistant to change, did not accept individual personal currency—but it did take currency from various "trusted" larger corporations. SoCAL Logging was one of those corporations. Luckily for Sherri, SoCAL accepted personal currencies in exchange for their own credits). The wallet app on her phone made this process pretty fast.

Her plane ticket purchased, she arrived in São Paolo, arrived at a local bed & breakfast, and called up a rideshare to take her to the beach. All of which she paid for in United Airlines "miles" (a strong currency around the world). She converted a few more of her own currency to get even more points, giving her a good amount of "working capital" to get her vacation started.

Then, a few days into her trip, something happened that Sherri didn't expect. Upon arriving in neighboring Argentina, she found most merchants didn't value United Airlines miles as much as the merchants in Brazil did. Everything cost far more, and some merchants didn't even accept the miles. Because "prices" were higher, and therefore disadvantageous to her, Sherri decided to exercise her freedom: to choose a form of exchange.

She logged into her wallet on her tablet computer and sold half of her airline miles to get British Petroleum (BP) dollars. With a little online research, she found that first and foremost, "oil money" was the trending form of exchange in Argentina. Second, she discovered that oil companies were usually willing to buy airline miles, since they had to set up a lot of air travel for their customers. Sure enough, someone from BP bought her airline miles within minutes of her offer, and suddenly, everything in Argentina was much, much cheaper for Sherri, and also highly advantageous to every Argentinian business that Sherri shopped with.

Sherri ended up having the time of her life, and learned even more about exchange and growing her business while in Argentina. She even got to perform a few haircuts, as people would ask her what she did for a living and she would offer her services. In turn people would pay her in local currencies, which she used for the remainder of her stay, and used her "oil money" to buy less localized services, like taxi rides.

Because of personal currencies, she was able to participate in the local economy in a way that was advantageous for everyone, and they allowed Sherri to make more direct connections with the businesses she encountered.

A CONSTANT SALE

Sherri is an example of someone with her own business; a freelancer of sorts. But many people work in larger organizations, on large teams and as a part of a management structure. How should they use personal currency?

Mike is a car mechanic that works for Midas, a large automobile repair chain, which has a special relationship set up with their employees regarding personal currency. Essentially, Midas buys Mike's personal currency in exchange for company currency. Their only stipulation is that Mike's currency be valued in increments of time—so, 1 "Mike point" equals one hour of Mike's time, and Mike is paid when the company buys his time (this is, as we mentioned earlier, how part-time employment already works—albeit in reverse order—this is simply more transparent form of the same thing). For every hour that Mike is punched-in at Midas, their computer system auto-redeems his points. In essence, Mike points are a promise of Mike's time, and Midas Gold Coins (their company currency) are units of their confidence in their employees' abilities and professionalism.

This brilliant system fosters a trusting and open relationship between Mike and his employer, and allows for scaleable rates as Mike or other employees grow within the company (i.e. every hour of Mike's time can be worth more Midas Gold Coins as Mike becomes a more senior mechanic). This system also helps Midas employees by increasing the reputation of their currency (since its use automatically increases its trust rating inside our fictional Personal Currency App).

If Mike wanted to, he could also create another personal currency for separate work, as long as it doesn't compete with Midas.

He brews his own beer at home and, due to its popularity at neighborhood barbecues and among friends, he decides to create BeerBucks, which supplements his income from Midas— and is a commodity-based, rather than time-based. He decides to value it at 1 BeerBuck per 4 fluid ounces of home-brewed beer. With his equipment (which produces five gallons of beer every two weeks) he predicts he can make about 1,280 ounces every month. Therefore he sets his monthly BeerBucks creation limit at 300. He starts out spending this currency with friends, and it quickly rotates to local businesses. Eventually people love his beer so much that the value of his BeerBucks becomes very high, and with the value surplus, he buys more equipment so he can make more. He's then able to create more BeerBucks, and therefore more value for himself and his family.

Now that we have tools like smartphone apps and the internet, Personal Currency can be so versatile, so easy to adapt to individual circumstances, that any person with a smartphone could use it.

And not only individual people—larger businesses can make great, profitable use of it as well.

CHAPTER FOURTEEN

TRANSITION

Modern customer retention programs are actually forms of currency; any agreement is a "currency." Businesses can creatively leverage these today.

USE YOUR SCRIP

Some things are so commonplace that their true nature often goes unnoticed. Scrip—company money—is already everywhere; most people just don't think of it as such. In the United States, especially, a common form of scrip is gift cards. Gift cards are like a local currency, but instead of being exclusive to an area they operate within a company. They are simply a way to consolidate and guarantee value (measured in dollars) inside a certain company, across time. *Time* is an important value in creatively designing new value-based agreements.

Another example of company money is loyalty programs. Loyalty points usually take the form of "miles," "stars," "punches," or "points." These are really just denominations of company value that have terms applied; terms that usually dictate how much of this value is equivalent to certain company products or benefits. Most companies, especially larger ones, are very specific about these amounts, usually making sure that these exchanges are advantageous to themselves. For example, airlines are careful to ensure that the "miles" (loyalty points) they pay out have such a low value that they require a customer to make many purchases before they have enough to redeem.

Coupons are another form of currency. A coupon that guarantees a certain amount off of a price is, itself, a store of value. It is worth exactly the dollar amount saved; it just has conditions for its use: you have to buy something else.

These various tools are easy to alter and take little time

or energy to implement—but are effective for incentivizing customers to buy things. Most retailers think that manipulating their customers is their only short-term option for generating better sales—but not necessarily so.

Currency is just an agreement, or rather, the vehicle for an agreement between parties. Currencies help people clearly define what they are going to get from each other. An easy way to start using personal currencies is to create a coupon, a loyalty program, or even just a sale (40% off!). Almost any stipulation on price, or programs that change prices (such as loyalty), can be mutually beneficial ways to not only generate sales but to strengthen relationships between parties. This is why coupons are called deals!

Yet, loyalty and many coupons are sometimes used to obscure costs. They are designed to make people feel like they are getting something better than they actually are. This does not build trust and rapport between people. It indicates a division; an invisible distinction, between the "business" and the "customer," rather than an agreement, a bond, between parties on equal footing (in truth, between the business composed of an individual person, and often a business of many people). Rebates are a common example of this, where retailers sell an item at full price that comes with a voucher to be sent back to the manufacturer in exchange for a certain amount of the purchase price. Many retailers make this process complicated and difficult in order to intentionally hamstring the process for customers—customers quickly discover that the effort required to get the rebate is more than they expected and often takes too long to be very relevant (by the time you get the check back in the mail, you might have forgotten you went through the process in the first place).

Using currency as a tool to *enhance* exchange benefits rather than exploit people is the ethos behind personal currency. Business does not need to be under-handed to be successful.

LOYALTY PROGRAMS

These were originally designed to keep customers "loyal"—to make repeat customers out of otherwise one-time transactors. Loyalty programs encourage business by typically promising a reward for buying products/services more than once. Punch cards from an ice cream store or a sandwich shop are common examples. Yet loyalty can do so much more. Imagine a loyalty program that gives customers access to certain products when they reach a certain "points" level. Or a gym membership that only gives you a rebate after you burn a certain number of calories there.

Urban Outfitters, a popular clothing company, hosted a fashion show that customers could be in if they reached a certain loyalty level. This fostered incredible participation, and made customers feel more connected to the company—so that it was no longer just a place where they bought their clothes, but was the place where their unique personal styles were valued, appreciated, and publicly "lifted up."
Another company to get creative was FoBoGro, a Washington, D.C. grocery store which adopted a strange but entertaining program. Rewards for the program included a date with the CEO, a chance to invent a deli sandwich (which gets on the menu for a week), and perhaps the most famous reward, 30 seconds to grab anything you want for free.

The most recent trend in customer loyalty is making the program more game-like. "Check-ins" have become a popular way to earn loyalty points, where customers get points for physically being in the store, or for other small actions, like posting a photo of themselves at the store on Facebook.

Loyalty programs all use currency. Those "points" that customers get for making purchases? Currency. Vouchers for products or discounts? Currency. A loyalty program itself is like a currency microcosm. It is a tiny economy where value is generated by some action (i.e. making a purchase) and then is redeemed (recycled) when enough of it is accumulated to receive a reward. This is a simplified but effective description of what currency is.

GIFT CARDS

This is a more direct form of valuing business currency, but works well for locking value into the company. These have become more creative recently. For example, shopping malls have introduced gift cards that customers can use at any of the stores in the mall. Groups of businesses in certain areas have joined together to create common gift cards as well, with mixed results.

People often do not redeem their gift cards. They forget about them. What if, however, those gift units were tradeable? If one store's value was directly exchangeable for another's, people might think of them as currency (acknowledging, of course, that if they were denominated in dollars, as they usually are, the gift cards might as well stay dollars rather than change into interchangeable company credit).

"Gift cards" can transform into something creative, just like loyalty can. Imagine gift cards that, when used, automatically donate a portion of their value to some kind of cause. Or a gift card for students that "unlocks" when their grades get good enough, or when their attendance is consistent.

Even better: imagine a gift card program for physical training that automatically bills customers if they don't exercise. So the money the customers pay is in exchange for being reminded and coached to exercise, and rewards them (by not charging them) if they do exercise.

Another idea is micro-loans, where a store releases some small amount (like $50) in credit to loyal customers, who can use it when needed and pay it back when convenient.

COUPONS ("OFFERS")

Almost every business, from large to small, has used coupons or something similar to promote sales. Coupons are vouchers for a certain amount off of a product's listed price. They are so common that many people, before buying nearly anything, at least check to see if a coupon for the item or service exists. Often it does, and some people are able to reduce highly-discounted things like grocery items into nearly no-cost shopping ventures.

Recently coupon trends have turned to "deep-discounting," with services like Groupon and Livingsocial dominating the space. These, however, are just more extreme takes on the entire concept of discounting—not actually very new. "Offers" were originally more personal; today's coupons hearken back to when deals were arranged between shopkeepers and their customers individually, in a process called "haggling."

Yet offers have become more dynamic recently, and there are a myriad of possibilities. "Deals" really are just forms of currency—since money is an agreement, one-time agreements are like fast-redeeming currency. They are locked in to specific purchases, perhaps for specific products or services, and often exist only to promote a transaction that would not have otherwise happened.

Coupon (deal) creation is very easy, and creative— imagine viral coupons that would not work unless the customer duplicates them and gives one to someone else, or discounts that start out very high and lose value over time (prompting customers to purchase quickly). Or deals that only work for certain people; a merchant is free to break their customers up into groups that can only get certain offers (like offers based on age; discounts on ice cream for children, or discounts on first-time cell phones for teens, or a discount for fist-bumping the CEO. The possibilities are countless).

"Transition" to monetizing your currency need not be any kind of ordeal—you can start doing it today. Notice that you do business with a few regulars? Give them a loyalty program, and make it creative. Want to encourage young people to shop at your store? Give points for social media posts made at your store, or hide QR codes around your business and challenge smartphone users to find them—and reward success.

Or create a company currency right away and integrate it into deals, promotions, loyalty programs, or give it out to customers as you see fit. You might be amazed at how personalized currency makes people feel connected to a business they use—how people naturally appreciate and unconsciously reciprocate when they feel genuinely served by a business.

Your business is you, and you are your business. Currency

is the "blood" of business, and should creatively, dynamically carry the agreements we make with each other. Create incredible opportunities for new relationships and true friendship with customers, who aren't only just that, but are fellow businesses whose value you are competing for. Give them equal footing at your place of business and they will, as is human nature, make a connection with why you do what you do (you know why you're doing what you're doing, right?).

Because personal currency will only make sense when created and used with sincerity.

Part IV: The Future of Currency

"I believe that banking institutions are more dangerous to our liberties than standing armies. Already they have raised up a moneyed aristocracy that has set the government at defiance. The issuing power of money should be taken away from the banks and restored to the people to whom it properly belongs."

Thomas Jefferson

CHAPTER FIFTEEN

THE TABLE OF COMMERCE

**The most anchoring obstacle to the progress of
currency is the practice of forced usury.**

AN UNEXPECTED GUEST

Four people sit at a table—you're one of them. Each of you brought a different kind of food to eat. You see that one of them has a delicious, white, fluffy cake. You want some—but the person who brought it isn't interested in what you have. They're looking at the salad that their neighbor brought. And that neighbor is looking at the pot roast your neighbor brought... and that neighbor is looking at your tacos. In this scenario, no one can barter directly. You don't have what the cake baker wants, nor does he have what the salad tosser wants, and so on and so forth. It's a predicament.

But then a fifth person sits down, and instead of food, he seems to have brought a stack of bills—cash.

"Here," he says, "I'll give you some of my money, and you can use it to pay the cake baker for his cake. Then he can use the same money to pay the salad tosser for his salad, then he can get some pot roast... you can all get what you want." That sounds like a good idea, and you're glad this fifth person sat down. He is helping you solve a problem.

"But what do you want in return?" The cake baker asks.

"A relevant question." The fifth person says, grinning. "In exchange for the ten dollars I'm going to loan your group, I just want an extra dollar on top of my original money in return."

Upon looking around, it seems that everyone considers this to be a fair price to ask. Just one extra dollar—ten percent. Thus the merry-go-round begins: first you pay the cake baker, who gives you his cake. Then the baker uses the money to pay the salad tosser, who forks over the salad. Then he pays the pot roaster for his pot roast, who then gives you the money for the tacos. You are about to pay the moneyman back, when suddenly, you realize you only have $10—but you owe him $11. As he watches you, a grin breaks across his face.

"How am I to pay you?" you ask. "I owe you interest, and yet, the money to pay it isn't here. Would you like a taco instead?"

The moneyman waves his hand and shakes his head. "No," he says, "We made this agreement using money. So I'd like money."

You look around, wondering what to do. Then the pot-roaster says to you: "Psst, listen! I'll just get a loan from him and I'll give you the dollar. Just pay me back when you can."

He does as promised, lending you the dollar. Grateful for the help, you hand it over, and your debt to the moneyman is paid. But what about your debt to the pot-roaster? You suppose that you'll get the money to pay him back sooner or later. The pot-roaster, on the other hand, now owes the moneyman $6, and only has $4. So he appeals to another member of the group. Each time someone borrows from the moneyman, he prints some more money so that he can lend it out.

In short order, a huge number of dollars, far more than were required originally, are circulating. Not to mention the moneyman is owed all of it, and everyone at the table is essentially working to pay him off, directly or indirectly (like in your case, you don't owe him directly; you owe your friend the pot-roaster, who owes the moneyman). The moneyman, meanwhile, just grins away, like he discovered a foolproof way to profit without actually bringing anything to the table.

Because that is exactly what he did.

Modern banking has become a way to get something from everyone, in exchange for nothing—and to keep doing so forever. The reason that the system is able to work without everyone being totally unable to pay the bank? New people keep joining.

Yes, indeed, our dollar-driven economies are run by banks, and the reason they keep going is that more money is being continuously lent into the system. And in this unfortunate reality, everyone is working for the bank, whether they mean to or not.

A Table of the Near Future

We have to admit, despite the clear discrepancy of value offered versus value given, that the "moneyman" solves a problem that the businesses at the table of commerce have.

Let's imagine, now, a slightly different scenario. The four people at the table are still joined by the fifth person—the banker. Money is still necessary for the operation of business between all of you, since direct trades can't always take place. But there's a key difference in this scenario.

All of you still use dollars, but one day you and another merchant, let's say the cake baker, decide that you don't want to use the moneyman's dollars as much anymore. "Gosh," you say, "Why don't we just create our own money not charge any interest on it?"

"Good idea," the cake baker agrees, and you begin creating your own money. However, despite your agreement, the pot-roaster seems totally against it—she prefers to use the moneyman's dollars, as they have always worked. So she flatly refuses.

The salad tosser, on the other hand, is open to your new proposition, but cautious. He states that he will accept your new currency in small amounts only.

"Perhaps," he says, as he eyes the moneyman, "up to 20% of our business could be done in your new currency."

This seems well enough to you and the cake-baker, and you strike a deal to create a certain number of units called "Noms" to be traded in this fashion.

In this way, you can reduce your dependency on the moneyman—even by only 20%. That means at least 20% of the money you use is interest-free, and the more you grow this percentage, the less dependent on the moneyman you will be.

A TABLE OF THE FARTHER FUTURE

Ultimately, though, you don't want to be dependent on the moneyman at all. While you do need a device for exchange, you don't need one that perpetuates dependence and debt on everyone who uses it.

So, a third scenario. Whether all at once or one by one, everyone abandons their dependence on the moneyman's "product" and instead you all agree to create your own mutual "community" money.

If you had a community money that helped you mutually clear debts between your businesses, you would have what is commonly called a LETS, or Local Exchange Trading System. Most LETS only work as a portion of the total money in the community—like in the scenario above—but they could, in theory, work as a full currency replacement.

Why stop there? With a few tweaks, you can further increase the currency's effectiveness and usability. Since each person at the table has a distinct separate business and distinct separate needs, the creation of currency could be more "localized." That is, each of you could create his own currency. This currency could be directly redeemable for your respective products, so that holding the currency is holding a direct representation of a valued product. Then, as each of you paid to receive another's product, you could be confident in the worthiness of your exchange device.

This type of system would increase trust between all of you, as the terms of your business would always be clear. While LETS are equally transparent, the LETS currency itself is of no value or substantiation outside of the LETS itself. Money based on commodity would be potentially valuable anywhere, and could instantly change to suit different circumstances.

Communities of the future will likely transition from a system of dependence on "The Moneyman" (the US Federal Reserve and its extended banking network) to a partially personalized currency system. Alternative devices for exchange are already being accepted everywhere, and continue to appear in the form of online applications and highly social peer-to-peer services.

This transition is not just a utopian hope or a call to action for a niche audience. It is inevitable. As the US dollar and other national currencies decline in their practical utility and their ability to reliably store value, combined with a continuing trend toward customized electronic exchange, community currencies will emerge out of pure necessity.

When faced with a problem, people always find a solution. In this case, with personal currency, a solution is just waiting to be implemented.

Chapter Sixteen

A Vision of the Future

Although it is a form of personal credit, Personal Currency could be called "resource currency"— and will circulate in the resource-based economies of the future.

It Was Tried Before

One of the problems with fiat currencies is they lack objective valuation—this is why it's easy to create a currency as long as its value is inherently dependent on the perceived value of something else. A common example is how many countries directly value their own national currencies on the US dollar. The problem is, what about the dollar? How does one determine its objective value?

Before currencies became truly "fiat," or given value only based on government decree, the world used a standard based on gold and silver. By using these commodities, currencies stayed stable and relatively transparent in practice.

The departure from gold/silver standards was actually a complex process. International currencies remained largely stable and suffered little inflation under these standards—their only major inherent flaw was that it made money dependent and therefore vulnerable to changes to the supply of these precious metals. For example, when gold was discovered in California in 1848, the US economy received a shock that benefited few—prices rose, due to the individual unit value of the dollar decreasing (since there was so much more gold, it was not worth as much by the ounce), yet this happened in such a short period of time that wages did not rise along with them. This made people poorer, and caused US exports to become more expensive, which discouraged trade with the US. Seeing a pattern? The sudden change in the supply of the key commodity creates a whole mess of economic problems. Mainly, the money supply could not be adjusted in a realistic timeframe to suit economic conditions, leading to things like "panics" and "total bankruptcy during wars." This condition ultimately led to the demise of the commodity standard, and the quiet adoption of fiat valuation.

Yet by the 1990s it was clear to the US government that the dollar was in need of some kind of objective value, so its purchasing power could be effectively compared to other currencies and to itself at different times. So in 1996, the Department of Commerce created the chained-dollar method of valuation, which was a way to value the dollar based on its purchasing power in any given year. This gave money an effective value that helped the government understand how much they should print in a given year.

But this kind of valuation method can only work for a huge currency, one that is used every day, by hundreds of thousands of people, like the dollar. Only then is there sufficient data for comparison.

And even then, in the end this currency is still an instrument of debt and little else.

"Buy Gold Now"

There has been a craze to revitalize the gold standard ever since its abolishment—which is an unlikely prospect. Our system of money and banking is now so hopelessly tied up in layer upon layer of twisted law and complex valuation that trying to drag it back to an old system would break more than it would fix. Indeed, it would not be a real solution anyway. We would bring back the same problems that plagued the monetary system before we departed from the gold standard, such as price fluctuations and increased possibility of runaway inflation or deflation. The subsequent increase in gold mining would devastate parts of the environment, since almost all of the easily-mined "surface" gold on the Earth has been found.

The parts of commodity currency that are useful, such as stability, transparency, and widespread recognition are worth keeping —it sort of forces money to stay tied up with something real. You can not mathematically or legally change the supply of an actual object, nor can any one authority decide how people feel about it in general.

Another useful feature of commodity currency is that it does not necessarily have to be based on a precious metal (as it traditionally was). A modern example of a dynamic commodity that helped give currency its value is oil, which gave rise to petrocurrency. Petrocurrency is not literal money, but rather a term for national money that is strongly influenced by the relative price of oil, such as the Canadian dollar, which usually rises or falls along with the value of oil (since Canada, as of this writing, exports so much oil). However, petrocurrency (that is, those national monies that rise and fall in value along with demand for oil) was and is not a true commodity currency. That's because the connection with oil that these currencies have is a phenomenon, and is basically accidental. Canadian dollars are in no way meant to reflect or indicate the price of oil, nor is the price of oil somehow dependent on Canadian dollars; they just happen to correlate because of unique economic conditions.

Businesses of the future are likely to use all kinds of commodity money, however—and actual, purposeful commodity money. In fact, companies that work in the same economic space will likely create common currencies that rely on the commodity that they depend on, though not necessarily deal with directly (such as how shipbuilders who build oil tankers do not deal with oil directly, but their oil-tanker production relies on demand for it).

A Resource-Based Economy

As money and our prevailing practical ideas surrounding it change, our economy will also change. Under a fiat money-based system, in which the underlying goal for every business is the accumulation of more money, demand must constantly increase, and therefore, so must production. This causes our economic system to produce an incredible amount of waste. It has to—when the entire goal is to make more sales, more than your competitors, you aren't concerned with conservation of resources, or energy efficiency (except whence it relates to your ability to make sales). An example of this frivolity is in the production of eating utensils—many different companies make things like forks, knives, and spoons, and all of them are competing with one another. This results in a huge number of excess utensils being produced for no reason other than to sell more of them.

The core idea of resource-based economy is that all of Earth's resources are the common heritage of humanity. Therefore we should divide and apportion these resources according to our need (not our greed) as a species. Indeed, the resources of our world are what are truly valuable to our human existence. Without resources, money itself is worthless. Yet we tend to treat money as a commodity; as if it holds weight and utility for our lives by itself.

This causes a massive disparity between need and provision. We have the production capacity and material to create enough food and essential products for every person on the planet; yet this capacity and these materials are instead being wasted to produce excess products, like more and more plastic food containers and enough cars for every person to have two.

In reality, we do not maximize the potential of the products and tools we do have. For every 45 minutes we use our cars, they spend 9 hours and 23 minutes parked, doing nothing. The cooking tools in our kitchens go unused for the bulk of the day as well. Imagine instead if families shared things like kitchen appliances, or their cars. Between them both, they would use half the materials and energy they did before.

Efficiency is a better use of existing resources, including time. With a few simple adjustments to our ways of thinking and operating in the world, we could make better use of what we already have, and thus prevent waste and poverty.

The actual meaning behind the term "economy" is very simple—to economize! To create more efficient processes and to use fewer resources. And yet, that's not what our "economy" does. Indeed, the free-market system actually prevents efficiency. That's because efficiency, in many cases, can reduce profit: the motivational force that drives the market. Paradoxically, in our modern "economy," poverty is an operational necessity.

Profit is the reason we have poverty. To have profit, there must be scarcity, so that there can be a relative force to "drive" people (the market) forward. Poverty is simply a more permanent form of scarcity, and profit is the way to escape scarcity. Thus, we have an economy built on escaping a problem that we ourselves have created.

Scarcity keeps people under control; it keeps people working to escape it. Modern money is a mechanism of control, which is one of the reasons we need the monetary system to change. We're in need of a new motivation mechanism; something for people to strive towards other than the accumulation of money that has a built-in disadvantageous contract; it's rigged.

By having a resource-based, rather than capital-based economy, we could create more efficiency. The "resources" that we need to use more efficiently don't just include things like minerals, water, and land, however: they also include people. Without having to unnecessarily strive for fiat money, individuals could instead efficiently use their time and energy to cultivate and grow useful and practical skills of their choosing (their respective businesses), thereby contributing to their communities in a healthy, authentic way.

Personal currency is a fully accountable way to count resources, whether they are commodities, raw materials, services, or time. They will naturally introduce a way of thinking that resource-based economy requires.

In other words, personal currency forces users to think of money and resources synonymously, which will greatly simplify fiscal management (since it will remove the disparity between the amount of money and the amount of actual resources available).

In the future, the management of resources will take on massive significance. The unsustainable nature of our current monetary paradigm will force us to begin economizing the most limited resources, or face crippling shortages. In fact, when a shortage does arise, some central authority usually steps in to manage that resource until its supply stabilizes. This is because the "free market" can not be trusted to effectively economize anything! So why wait until resources, especially the essential ones like potable water, are running low before we start managing them? We could be "economizing" right now, but instead, we're wantonly grinding trees into cigarette filters and toilet paper, transmuting oil into more tupperware than anyone could ever use, and dumping drinking water onto lawns to keep them green.

However, resource management need not be completely centralized—that is to say, rather than some kind of governmental authority managing resources, in large part, we could instead manage our resources individually, and on a family basis (small-scale management, or SSM). We just need incentives to do so.

We need to change the mechanisms which shape our thoughts about economization—much like how our opinion of traveling has changed according to our technology (because of planes, for example, traveling across the country is not met with nearly as much hesitation and dread as it used to be), our thinking about resource management is continuously confined by how our monetary system works. We need a monetary system that corresponds to natural laws, to the natural restrictions of our existence on Earth. Because all individuals take up space and energy on this planet, it's only natural, then, that we all take part in the management of resources.

CHAPTER SEVENTEEN

SYSTEMS THEORY

Everything, directly or indirectly, affects everything else. Harmony exists when resources flow fluidly on the paths of least resistance. Effective resource management means reducing obstacles to this flow.

TRUE RESOURCE MANAGEMENT

Household management is common to every family, regardless of nationality. And, as was mentioned in Chapter 3, this is the origin of the word "Economy:" oikonomia. The efficient use of resources is actually built-in to our social structure and our family culture, and has worked since humans first walked the Earth. From our forest and grass-dwelling days to now, we have had to consider how to use the limited resources available to survive, to thrive, and to plan for later. The concept of "resources" is built in to our minds.

Yet we are entering a new and unforeseen era. We don't have the problem that some peoples of the past had; that the land we occupied was becoming unable to sustain us, and the solution was simply to move on to a different land. Our species covers the Earth, and can't do that anymore. There's nowhere else for us to go—we have to make this work. We need an effective resource management system.

This is a well-known fact, however. It's no secret that a massive change is needed in the way we manage resources. The thing that no one can seem to agree on is how. What kind of system are we supposed to use? How can we use something that works for everyone?

To solve the problem of economy, we have to go to the basics—to natural systems.

WE ARE AN OCEAN

No matter your size or individual needs, your territory, or your species; if you're an ocean organism, you're connected and interdependent with everything else around you. The Earth's oceans are replete with complex interdependent relationships, such as that between whales and krill, some of the largest and smallest organisms in the ocean, respectively. Baleen whales (such as the humpback) eat krill, which feed on plankton, which feed on waste... such as that from whales, thus forming a circular resource-based energy relationship. Other organisms have directly mutual relationships, like that of clownfish and sea anemones. Clownfish dwell among the anemones' tentacles, which deter predators by stinging. The clownfish, in turn, clean the anemone of parasites.

Systems Theory is the idea that everything is simultaneously dependent on, and depended on (whether directly or indirectly), by other things, and these relationships are the very fabric that makes up the universe. Systems Theory is especially noticeable in living ecosystems, where the very existence of organisms is predicated on these relationships.

Any human economy can be described as exactly that—an interdependent web of relationships. We have mutual economic relationships, and we have more distantly circular relationships, like those formed by supply chains. And, as many political activists will eagerly tell you, we have parasitic ones. And this is where understanding Systems Theory becomes beneficial.

In ecosystems, parasites recycle energy back into the environment, and act as food for some creatures, often called "cleaners," who keep their populations in check. Cleaners developed in response to parasites, and ecosystems with cleaners are more efficient than those without.

But in human economies, parasitic entities are not part of the energy cycle. When the very vehicle for energy is engineered to return to and coalesce in certain entities, then the whole ecosystem ends up lopsided, causing organisms that are less parasitic to be stunted at best, and at worst, starve and eventually die.

This, of course, is a metaphor for bankruptcy (or insolvency)—an eventuality in our centralized, fractional-reserve monetary system.

In nature, more efficient systems continuously replace less efficient ones, and organisms adapt to changing circumstances, in a process known as evolution. Human society has also evolved, becoming more efficient and adapting to challenges continuously. And we, just as natural systems and organisms often do, have encountered a new challenge we must adapt to... but this one is different.

This adaptation is to a man-made problem: the inefficiency of our existing monetary paradigm.

A Systems-Theory-Based Solution

Understanding how the economy operates is the first step to finding a solution. How can we implement Personal Currency without relying on US dollars, or on some other outside system to stabilize it?

Again, we can look to nature. At its base, all life on Earth is made up of systems that operate in simultaneous independence and dependence on one another; in other words, they are interdependent, while also sovereign in their uniqueness. The only "centralization" we see is a result of emergence; in other words, these different interdependent systems regulate one another through mutual feedback.

In ecosystems, all organisms trade common resources— energy and chemicals. And these common resources act just as money does in our economies. The key difference, and the reason that we haven't had a truly eco-systemic free economy yet, is that money has almost always originated with central authorities, whether they were banks or governments. No one but them was able to set the terms of monetary usage, and as a result, the economy was always lopsided to favor them.

The only way that we can have interdependent yet sovereign businesses who together make up a truly free economy is if we decentralize money. For the first time, this is possible through digital technology; through the Internet.

A "Digital Reserve" for each person, for each business, will be readily possible on personal electronics like phones and computers, reflecting and representing the potential energy and expertise (resources) already present (in reserve, so to speak) within those people; those businesses.

With decentralized money, we can finally experience self-regulating, "natural" economy. And now that we have the technology to create it, we can start doing it right now.

A mobile application platform using cross-platform technologies, equipped to allow different applications to run in tandem or to exchange data in real-time, is not only doable, but is not even difficult with today's development tools.

A digital exchange platform won't fully replace money. But used alongside money, tools like an application platform can gradually lessen our dependence on it. To begin, businesses can use it to keep wealth within communities. As time goes forward and devices like the US dollar become less and less stable, personal currency will naturally grow in popularity as it proves to be a viable alternative.

And we need a viable alternative. Soon.

CHAPTER EIGHTEEN

A TIMELINE FOR CHANGE

History is a chronicle of social change. Everything must evolve into its next natural, superior form. We have little problem applying this to technology—but societal change is not implemented due to the inefficiencies of the social structure we live in.

CURRENCY ALTERNATIVES

Let's review our current monetary model.

The US dollar is created by banks, including the Federal Reserve Bank, when someone borrows from them. These instruments of debt must then be paid back to the banks with interest; however, there is not enough money in existence to pay the interest, which forces the constant creation of more money, and therefore more debt. This leads to economic authorities promoting growth above all, and success or failure is measured in dollar profits (or lack thereof).

The only way that we can trade outside of the debt-credit system is to barter, or to exchange our debt credits for company gift units (scrip), like Target or Best Buy gift cards. Even these instruments of exchange are inseparably attached to the debt credits we use, however.

We aren't able to trade with anything else because our entire economic system is built on capital, which is another word for money. The entire goal is to gain more of it, and the consolidation of the concept of wealth within money is therefore necessary for this to work.

It's unsustainable. It won't work in the long-term, and it's already shaking where it stands.

The next natural step in a currency reform is gradual adoption of alternative methods. Because the world's economies are so mired in dependence on debt money, adoption has to be gradual. We can't just drop everything and begin using personal currency all at once. But we can start to use different methods for exchange alongside our current devices. In fact, we already do so.

Every day, we exchange without money, without thinking about it. We trade chores, tasks, compliments, time, and more. These things are usually between individuals or in small groups, but they operate alongside money nonetheless. In fact, many of us prefer the feeling of informal trading; it feels more personal than using money!

When people begin using personal currency, they will naturally begin to understand that value is a flexible notion. Much like how your opinion of different foods can change as you get older, and you even begin to see value in concepts such as nutrition (something which you were not born knowing), your perception of money, and what makes it valuable, can and will change.

But this change will be gradual, as long as the US dollar exists. People will feel more comfortable comparing upstart currencies to the dollar, since it has been such a constant their entire lives. Personal currency will begin its valuation directly in dollars; much like the alternative currencies that already circulate.

There's an important element to note about this change, as well: it will begin with the middle class, and travel outward from them. They have disposable income—in fact, it's the middle class that drives the American economy, since their spending habits are, cumulatively, the largest—and this disposable income means that they are able to convert some of their money without losing everything. This is certainly true of the "upper" class too, but in their minds, they will have less to gain, since they are already independently wealthy. Not to mention that those with a very high income are less inclined to advocate, or participate in, a new system of currency which gives more purchasing power, and therefore more economic advantage, to more people. The middle class, which is increasingly difficult to stay in and is now almost impossible to move up from, will have little reason not to try creating their own, unique, interest-free and debt-free wealth. The "common" people are the ones who actually do jobs; they manufacture products and perform the services of the economy. Practically, the middle class *are* the means of production in the economy, and currency, used properly as a reflection of the economy, should reflect this. In other words, currency should be personal, direct, and substantial—just like the people who create/own it.

THE END OF FIAT CURRENCY

Though perhaps difficult to imagine as of this writing, there was a time, not so long ago, when money was diverse, disparate, and uncommon. Yet economies thrived and trade abounded, with fewer restrictions and, it could be argued that business was much easier to do. Human economies have grown in size since then, but the basic elements therein are still the same. Resources, production, workers, management—it's all there. A world in which money is decentralized is perfectly doable. All we have to do is adopt updated ideas of what that means. Just as our cultural ideas about other seemingly established and unchangeable things, like race, have evolved, so too can our cultural ideas about what money is.

When money was converted to a group of one-dimensional fiat currencies it became both consolidated and much, much more complicated. The complexity was intentional, to control its flow and ultimate destination. But by diversifying and simplifying money, personal currencies will quickly supplant and overtake fiat money; therefore it will be defeated not by legislation but by obsolescence and subsequent disuse.

Before this even occurs, we will see an explosion in new businesses, and highly localized economies will quickly stabilize, since value does not need to travel as far as in places with long supply chains (like in some cities, or in industries like technology or manufacturing, where goods are sourced and assembled from all over the world).

Unfortunately, however, there will be negative consequences to ending fiat money. Governments will likely do anything within their power to prevent their currencies from being supplanted by decentralized currency. The only reason the United States of America is able to wage endless wars, buy off foreign governments, pursue inefficient and frivolous projects, and support multinational corporations with impunity is because they control the most coveted and dominant currency in the world, and are able to make as much of it as they want. They need not fear any regulation—since they are the regulators. Why would they want such a system to end? In a decentralized currency system, they will be forced to account for every single transaction they make, and will also need to make investments in other personal currencies.

Governments around the world use fiat money to levy hidden taxes, control business, for the military exploitation of weaker nations, and to protect the unregulated growth and manipulation of the insane, useless banking industry.

A new currency paradigm will do much more than just empower businesses and free people from debt. It will actually erode the government stranglehold on money, and therefore businesses, everywhere.

CHAPTER NINETEEN

A DIGITAL RESERVE

How a master ledger of every personal currency could be decentralized and managed transparently.

WHY BITCOIN TOOK OFF

As the first widespread digital currency, Bitcoin accomplished a huge milestone. It did, after all, popularize (in relative terms) the idea that money could be digitally created, and it is the first practical currency in recent history to be fully decentralized and transparent. These aspects of bitcoin serve as a model for other digital currencies.

Bitcoin uses what amounts to an online spreadsheet that keeps track of every transaction ever made with the currency, and everyone's account is viewable, but anonymous. The peer-to-peer network, which is a set of computers that independently record and calculate the huge bitcoin "ledger" composed of anonymous addresses and their bitcoin balances, as well as a record of all the transactions ever made.

What this means is that by principle, nothing can be hidden, and therefore no single entity or organization can covertly control the system, nor can hidden taxes be levied on holders of bitcoin. This was one of the reasons for creating bitcoin in the first place, and is a good way to add built-in anti-exploitation measures. A decentralized digital reserve could implement the same structure and technologies to easily ensure the transparency of the network.

Bitcoins could be used for online transactions more securely than traditional money ever could, and for this reason, bitcoin is a good model to consider when creating new digital currencies.

New systems could build upon the same concept, making currencies that are simpler to store and easier to transact with.

FASTER, BETTER, & MORE EFFICIENT

There was another reason for Bitcoin: creating money that could do the same things, and more, much faster than the dollar. Free from cumbersome digital banking fees and absent the complicated, dense, and nebulous financial systems constructed to keep fiat money flowing down certain channels. While this is true in principle, however, in practice bitcoin is very complicated. Its creation process, which involves bitcoin users updating the chronologically-organized "block chain" (ledger) by using their computers to solve a cryptographic puzzle, is difficult to understand for the average user. As the bitcoin creation process becomes more difficult, only those few "elite" users who can design and build "mining" computers, which are specialized processing machines that mine bitcoins faster than more generalized PCs.

Digital currency, like bitcoin, has the distinct advantage of being able to be designed from the ground-up to meet the challenges of the modern world. One of the main challenges is security. When you pay using a credit card, you give the merchant (the person charging you) all the information they need to take your money away. This is how many identity thefts and credit card abuses occur; people pay using unsecured systems or they pay untrustworthy parties who then steal or sell their information. The same goes for bank accounts, which can be compromised (irreversibly) the same way. Because the systems we use to pay with dollars are all based on technologies invented decades ago, we are still dealing with these security issues today.

Instead, payment information should only allow someone to receive the amount of money specified by the payer. Digital currencies can be constructed this way from the ground up instead of having to be retroactively meddled with to make them secure. There were not any secure, nor sufficiently complicated, internet frameworks that could handle this kind of thing. That's why it's important that we begin work on these currency systems now, while the technology exists but governments haven't harnessed and controlled it completely yet.

Bitcoin also introduced the first self-sufficient digital money. In other words, it doesn't rely on a third party to keep your funds. With dollars, if the third party that holds the funds (a bank, in other words) goes out of business, you lose all your money. Although there's now federal insurance that mitigates this problem, the system is flawed at its root: the very fact that the money is held by a third party is anachronistic and unnecessary. This enduring aspect of bitcoin made it somewhat popular as an investment.

With a public ledger maintained by all users, this problem is nonexistent. Since everyone has an open record of the digital currency, if one entity "goes down," the system's integrity is maintained (and continues as long as there's at least one user).

Computer encryption and non-tangible money also means that theft is much more difficult, and individual transactions are technically safer, since the cryptographic keys used for transactions work only one-way and don't grant any access to digital account controls.

The drawbacks to bitcoin, however, are substantial. For one, the creation process is highly complicated and not understood by the average user. Payments made using bitcoin also require learning about how the wallet and creation processes work and understanding, at least at a conceptual level, how the cryptography at bitcoin's core works. It's almost as if the creators of bitcoin wanted the currency to stay within a technocratic community, rather than being used by a wide variety of people, or in mainstream society.

Another reason is that bitcoin tends to be unpredictable. Its value fluctuates wildly at times, sometimes simply because of press (both good and bad). And, of course, the real problem is that bitcoin's perceived value is based entirely in United States dollars. So how could it be a replacement for such if it relies on it for valuation? In the design of any new medium of exchange, using the U.S. dollar as a reserve currency is a fundamental flaw and a self-defeating idea.

We need digital currencies that can not only substitute the existing features of money, but truly exceed them, proving more flexible and useful.

Thus, future digital currencies will need to do what dollars do, but much faster and with much greater security and ease of use. Similarly, cash will become an unnecessary concept when we have digital tools that are just as convenient and fast. Otherwise, new currencies will just be passing fads, while fiat currency continues to hold us hostage to unnecessary institutions.

The primary features that digital currencies should take away from bitcoin are:

1. **Decentralized;** created and managed by all.

2. **Transparent**: has a meta-ledger, viewable by everyone.

3. **Anonymous transactions**

4. **Abstract storage;** can't be destroyed or seized without consent.

5. **Built-in cryptographic security**

6. **Community-sourced**; created by users rather than a bank.

7. **Difficult to tax;** clever abstraction means it is hard to define for tax purposes.

Characteristics to avoid:

1. Difficult to understand for the average user.

2. Highly technical creation process (favoring those who can build special hardware).

3. Valuation in government currencies (U.S. dollar, Euro, etc.).

4. Learning curve to adopt and use.

Overall, bitcoin was an exciting harbinger and a pioneer for digital currency. It opened the world's eyes to the possibility of fully digital money, and made a significant cultural impact that has begun the shift in perspective that is required to reform money. Now, it is our job to take what bitcoin established and create better, more secure, easier to use, and fully independent forms of digital exchange. With improving technologies and a greater awareness of the pitfalls of fiat money and digital security, we can make this happen.

CHAPTER TWENTY

DESIGNING THE FUTURE

The world needs a currency revolution, and the market is ready for personal currency tools. Who will make them?

PALE DEATH SITS AT A DESK

One thing that the creators of almost every alternative American currency in the past 50 years sought to do was bring more freedom and control over buying and selling to currency users. And they all had this in common: not enough people used them. Few were fully self-sufficient, which was their major flaw from the beginning—they relied on the dollar for valuation. As a result, these currencies failed to make any significant change to monetary customs in the United States. Thus, their common desires were never fulfilled.

The importance, and subsequent reliance that our economy has developed for our monetary system has taken on epic and disturbing proportions. Currency was once a far more fluid instrument, taking many forms that were interchangeable and carrying different connotations and uses depending on the situation. Now, money has become a one-dimensional instrument controlled by the same people who spy on their own citizens and stage military invasions of foreign nations under false pretenses. Rather than giving the power of setting exchange terms to the people making exchanges, we rely on political entities with zero economic output.

Democracy has one idea at its core: that people who live and work in a society should have the power to determine their own way of life. Democracy is as much economic as it is political, as well; since we are all businesses (what are you busy doing?), the freedom to do business as you feel called is as relevant as the freedom to believe and say what you will.

Democracy is fundamentally opposed to the idea of aristocracy, which holds that only certain individuals have the capacity, and should therefore have the power, to rule over others. In a monarchy, those aristocratic individuals are royalty, often determined by nothing more than lineage. In a dictatorship, the aristocrats are those with martial assets (read: guns and soldiers), and have authority only because their people fear death and pain at their hands. Any and all of these forms of government are authoritarian; thus, fundamentally incompatible and opposite of democracy.

The most powerful and dangerous aristocracy of the 21st century—the one which stands as the single biggest obstacle to economic progress and individual freedom—is not based on monarchic ideas of blessed lineage, nor is it based on direct martial rule; forms we are familiar with and exist in various places around the world. The aristocracy we now live under is a plutocracy: a system of rule based on money, and the manipulation thereof.

Political power and dollars are now synonymous. Yet it isn't a successful business baron who rules over us. It is the opposite. The creators of money are not businessmen at all; in fact, without legitimate businesses, they could not survive, because they do not produce anything. Through a slow erosion of understanding and political manipulation over the centuries, they have assumed a position of power over the actual businesses of the world. If they desired, could rend the billionaires of the world beggars with a pen stroke. The most powerful institution on this Earth, as of this writing, is not some political entity such as Congress or the European Union. It's not a person, such as the President of the United States. It's the United States Federal Reserve System.

When money was substantial and diverse, an entity like a state bank was just another bank; a "money-holding company." State banks had existed many times before, even in the US; yet when the Federal Reserve was created in 1913, it was new, bold, and vicious. Substantial money was subsequently abolished and wealth was consolidated into a baseless "reserve note" whose usage was enforced not by people's desire to use it—nor by the intrinsic value of products or services—but by bullets. In short order it became the single most powerful and influential institution in existence.

Entire nations fall and rise with the simple appropriation of the Fed's dollars and cents. Men and women with extraordinary talents and capacities for creativity will spend their lives in repetitive toil to gain a laughable sliver of what the Fed creates in a day, with nothing more than computers and ledgers. A currency like Bitcoin may indeed be composed of only binary code and nothing more—but at least it's impossible for anyone to control. The US dollar, the Euro, the Yen, Yuan, and every other state-sponsored currency have centralized control at their very center of existence.

At this very moment there is a small group of people deliberating over how much money to create today, and their decision will resonate through billions of lives, resulting in chains of circumstances that produce prosperity and plenty for some, but despair and destitution for many others.

Money Is...

More than just a device for moving goods and services. Money is like blood; it carries life through our mutual ecosystem; the vehicle for growth and prosperity. Its abundance provides security and the freedom to pursue a lifestyle of one's choosing. In our "capitalist" society, it is the "capital"—the very impetus of our economy, and therefore, our businesses. It is not to be treated lightly, and yet, its modern implementation is an insidious form of debt-slavery, rather than a fluid, substantial, and transparent instrument of value.

Without money, our civilization would not exist as we know it today. Money is an amazing invention, a piece of technology; one with great cultural significance and physical importance. It powered human progress for centuries. From small tokens and knickknacks, money grew and changed into a diverse tool for capturing value, and eventually, into a physical representation of agreements.

Agreements exist at the core of human relationships. They allow us to collaborate, to form workable patterns that organize and make us more effective. Because each of us is a business, it means these patterns form what we call the "economy," which is, therefore, just a web… of agreements!

These agreements are meant to be between and for individual businesses, and yet, modern money doesn't carry this convention. Currently, any agreement you make with fiat money ends up involving the institution that created it. Since we must all follow the stipulations that entities like the US Federal Reserve System and the European Central Bank (to name two), every single agreement made with their "reserve notes" includes their hidden terms.

Money could be so much more than that. It could be a tool for positive change; for favorable, healthy, and trustworthy relationships. It has the potential, even, to change the way we do our business—to enable us to do more, with less risk, and without having to wait for others to pay us to do our work. Money can, once again, be personal and substantial. It can be a means for empowerment rather than subjugation. It can become a cultural tool, rather than a cultural focus. We should, for our individual and for our corporate well-being, have money that has no power of its own; but rather symbolizes our relationships, which are the real sources of power in our economy.

Remember: each of us, everyone, is a business. We all have something that we are busy doing. Size is irrelevant—each of us has the characteristics of a business. Each of us requires capital to operate, employs individuals to do so (even if that is only one individual), and provides a product or service to others.

By contrast, businesses also take the form of amalgamations of individual human drives and gifts, and these are what we call "corporations" and "companies." They are no different, just larger and capable of more. They are still made of people, who contribute their individual "businesses" to the whole. Without these individuals, the larger corporation (a word meaning "body," as in, the combination of many interdependent parts) would not exist.

The ability to trade using money has allowed businesses to grow into the amazing working machines they are today, and also helped create the complicated networks of larger business relationships that produce the remarkable goods and services in today's economy. The very idea of money first spawned from the need to exchange between businesses. Since they are the basic components our society is built on, money is at the very core of life in the modern world; just as blood is at the very core of preserving physical life.

CREATING A REVOLUTION

Appealing to governments for change to monetary policy is not a tenable solution. Even if the political will existed, there is already a Bible-like tax code (which is actually about 50 times longer than the Bible) that serves as the bedrock of the US national budget. This budget is what allows every public agency and project to operate, from the Postal Service to the Department of Transportation, to every branch of the Armed Forces. Taxes are an indispensable part of the fiat system, because they go towards payment of the interest on the principal (the money being created). Without this façade that the debts are being paid, international confidence in the dollar would break down.

Our governments simply have too much too lose by disbanding or otherwise drastically modifying their fractional-reserve banking systems. Suddenly shifting to a substantial currency would force them to radically change their budgets or face destabilization. Their operations are essentially dependent on banking.

Social change regarding money, like most social change, must come from businesses. The mass production of automobile technology and a transportation revolution came from businesses. Revolutionary new ways to grow food using fewer resources and less space, helping feed more people than ever before was pioneered by farmers alliances (which are businesses). Incredible computer technologies, which doubled, tripled, quintupled computing power and plummeted in cost—making computers available to almost everyone—came from businesses like Xerox, Apple, Oracle, Sun, and Microsoft.

In fact, other than overly complicated taxation systems and national parks, one of the only significant modern proliferations that government entities have pushed into the human world is nuclear weapons; the most destructive and horrifying instruments of death ever conceived. Almost any technology invented by people paid by a government had either a direct or an ancillary military role—still, businesses manage to turn them into advantageous and appropriate consumer products and services, if allowed. Engineers working for Nazi Germany may have invented the first jet aircraft, but it was businesses like Boeing, partnered with airlines like Lufthansa that made jet air travel a common civilian convenience that the entire world benefited from.

Positive practicality gives anything, including technology, momentum to move through and past obstacles that invariably arise. That's because businesses, unlike governments, exist in the interest of creation rather than destruction. For any business, here's nothing else to fall back on but inspiration and hard work. A business can't just raise taxes or steal property if it's managing itself poorly. It has to adapt; to find better, more efficient, and less energy-intensive ways to do things. If it can't adapt and become better than before, it will die, much like an organism faced with new environmental challenges.

The future is in our hands, and always was. We must choose, as working businesses, to adopt more effective methods of operation… and there are few operational considerations more important than money.

INVENTORS OF TOMORROW

In the past, people wanted to be artists and engineers. Now, they want to develop software. Why? Because software, and specifically, mobile apps, are the newest and most promising tools of human infrastructure. While roads and waterways move materials, products, and people, applications that use the Internet move ideas, feelings, and currency between people. Communication is a key item in our world, and is where growth seems to be happening (and has few foreseeable limits).

Ultimately, tools for communication are most enduring when adopted by businesses. What are you busy doing? This is the essential life question (since we are all businesses), and when you know what your business is, you have to set about utilizing the tools you need for it. The faster and more efficient those tools allow you to be, the more successful and profitable your business can be—since communication is the key to gaining customers and partners in business.

Software is the Internet Frontier: a place where we can create new tools, yes, but also revolutionize old ones; like Uber did for urban transportation, or Mapquest and Google Maps did for consumer GPS. Software has the power to radically change our everyday lives, and that is why it is the perfect medium for the builders and planners of tomorrow.

We have a lot to gain by collaborating to design the next era of commerce. There are thousands like you out there; people wanting to make tools that have meaning and significance, not just a way to profit.

Yet there's plenty of that to be had, as well. The people who come in to the personal currency market first will have the strongest currencies in the long-term.

The long-hailed "Digital Revolution" is already here. We have the tools, the resources, and the people. All that's lacking is the awareness, and the will, to move.

Help build a currency community—and a better world.

"Watch money. Money is the barometer of a society's virtue. When you see that trading is done, not by consent, but by compulsion—when you see that in order to produce, you need to obtain permission from men who produce nothing—when you see that money is flowing to those who deal, not in goods, but in favors—when you see that men get richer by graft and by pull than by work, and your laws don't protect you against them, but protect them against you—when you see corruption being rewarded and honesty becoming a self-sacrifice—you may know that your society is doomed. Money is so noble a medium that it does not compete with guns and it does not make terms with brutality."

Ayn Rand

FURTHER READING

HISTORICAL WRITINGS

- *Forgotten Founders*, by Bruce E. Johansen.
- *Encyclopedia of the Haudenosaunee*, by By Bruce E. Johansen & Barbara Alice Mann.
- *The Southern Route "Out of Africa": Evidence for an Early Expansion of Modern Humans into Arabia*, by Simon J. Armitage, Sabah A. Jasim, Anthony E. Marks, Adrian G. Parker, Vitaly I. Usik, and Hans-Peter Uerpmann.

MONEY AND HOW IT WORKS

- *The End of Money and the Future of Civilization*, by Thomas Greco Jr.
- *Money: A Study of the Theory of the Medium of Exchange,* by David Kinley.
- *The Gift Economy*, by David J. Cheal.
- *Modern Money Mechanics*, by the Federal Reserve Bank of Chicago.
- *Money Creation in the Modern Economy* (Quarterly Bulletin 2014 Q1), by the Bank of England.
- *Paper Money of the United States*, 18th Edition, by The Coin & Currency Institute, Incorporated.

MISCELLANEOUS

- [A Firsthand Account of the Battle of B-R5RB]: **http://community.eveonline.com/news/dev-blogs/the-bloodbath-of-b-r5rb/**
- *A Theory of Human Motivation*, by Abraham Maslow.
- *A Cognitive-Systemic Reconstruction of Maslow's Theory of Self-Actualization*, by Francis Heylighen.

9 780989 228527